Quantitative Aspects of Science and Technology

Consulting Editor

Samuel L. Oppenheimer
Ohio Technical College

BILL G. ALDRIDGE
Florissant Valley Community College

Quantitative Aspects of Science and Technology

An Introduction

Charles E. Merrill Books, Inc., Columbus, Ohio

Copyright © 1967, by CHARLES E. MERRILL BOOKS, INC., Columbus, Ohio. All rights reserved. No part of this book may be reproduced in any form, by mimeograph or any other means, without the written permission of the publisher.

Library of Congress Catalog Card Number: 67-22016

Printed in the United States of America

1 2 3 4 5 6 7 8 9 10-75 74 73 72 71 70 69 68 67

To ALICE
for her help and patience,
and to BRYAN, CARA, and RICHARD
for learning forbearance when so young.

Preface

This text has been written for use in an introductory course in science and technology. It may be the primary book, or it may serve as a supplementary text for students enrolled in physics, electronics, chemistry, or related courses requiring technical calculations.

In writing this book, the author has drawn upon his teaching experiences with students in various disciplines. It is felt that the instructor whose students use this book as a supplement will have more opportunity to devote lecture time to actual course content, instead of being continuously side-tracked by problems concerning mathematical methods. The student will benefit in that he may take advantage of his full potential, gaining a thorough understanding of the *quantitative aspects* of technical disciplines, and utilizing these aspects in his study.

A number of students have difficulty with their science and engineering courses, not because they lack aptitude, but because, for whatever reason, they have never acquired certain fundamental skills needed to *begin* studying courses of a technical nature. Because students lack these prerequisite skills, they quickly become lost, or they spend excessive time just trying to keep up. The areas in which they need skills are: basic computation, including the use of a slide rule and power of ten notation; data collection and analysis; graphical construction and interpretation; quantitative problem analysis; and location and *reading* of quantitative materials.

The comprehensive coverage of common "stumbling blocks" is presented logically and, hopefully, clearly. Each topic has been selected to familiarize the student with the calculations most frequently necessary in pursuing further study.

Chapters 1 and 2 cover the definition of units, leading into a discussion of numerical symbols and operations. Because power of ten notation is so essential, it is given separate and intensive treatment in Chapter 3. The slide rule is introduced to the extent of explaining the C and D scales; the student is also acquainted with the necessary operations for calculating squares and square roots of numbers.

Chapter 5 covers the solving of algebraic equations, and Chapter 6 explains ratio and proportion, using illustrative examples. The measurement of fundamental physical quantities is dealt with in Chapter 7. Chapter 8 acquaints the reader with experimental data, and introduces some elementary statistics. Graphical analysis of experimental data receives attention in Chapter 9, and in Chapter 10 the student is familiarized with the procedures of the scientific experiment. Chapters 11, 12, and 13 provide information as to proper sources and methods of obtaining scientific and technical materials, and numerous examples of applications of problem solving.

The general plan of this book, then, is to proceed from the most elementary level of the basic quantitative skills to more advanced calculations. Topics are presented and developed inductively.

It is the author's sincere wish that those who use this text will attain their full potential in scientific and technical courses, becoming familiar with a body of knowledge sufficient to allow them to master related subjects in their education.

I would like to express my appreciation to the students and colleagues who made this work possible. Chapter-opening illustrations are used with permission from the publication *On Your Own in College*, by W. Resnick and D. Heller, Charles E. Merrill Books, Inc., 1963

Bill G. Aldridge

June, 1967

Contents

Chapter 1
Physical Quantities and Definitions   1

    1-1   Fundamental Physical Quantities   2
    1-2   Physical Definitions and the Operational Process   4
    1-3   Mathematics in Science and Engineering   5
    1-4   Units for Fundamental Physical Quantities   6

Chapter 2
Numerical Symbols and Operations   9

    2-1   Natural Numbers   10
    2-2   Integers   13
    2-3   Rational Numbers   14
    2-4   Addition of Rational Numbers   16
    2-5   Subtraction of Rational Numbers   19
    2-6   Multiplication of Rational Numbers   20

2–7	Division of Rational Numbers	21
2–8	Summary of Rules of Operation on Rational Numbers	22
2–9	Irrational Numbers	23

Chapter 3
Computation with Power of Ten Notation — 25

3–1	Powers of Ten	26
3–2	Powers of Ten in Multiplication	28
3–3	Powers of Ten in Division	29
3–4	The Negative Exponent	31
3–5	The Zero Exponent	33
3–6	Expressing Numbers in Power of Ten Form	34
3–7	Multiplication of Large and Small Numbers	37
3–8	Division of Large and Small Numbers	39
3–9	Roots and Powers	41
3–10	General Computations	42

Chapter 4
Slide Rule Computation — 44

4–1	The Slide Rule	44
4–2	Slide Rule Scales	45
4–3	Reading the Scales	46
4–4	Multiplication	51
4–5	Division	54
4–6	Multiplications and Divisions	57
4–7	Squares and Square Roots	59

Chapter 5
Solving Algebraic Equations — 64

5–1	Algebraic Symbols	64
5–2	Changing Verbal Statements to Symbolic Form	67
5–3	Mathematical Operators	70
5–4	Mathematical Equations	71
5–5	Solving Algebraic Equations	72

Chapter 6
Introduction to Ratio and Proportion — 78
- 6–1 The Meaning of Ratio — 79
- 6–2 The Meaning of Proportionality — 82
- 6–3 The Proportionality Constant — 85
- 6–4 Ratio and Proportion in Similar Triangles — 89
- 6–5 Examples of Ratio and Proportion from Science and Technology — 96
- 6–6 General Ideas of Ratio and Proportion — 100

Chapter 7
Measurements of Physical Quantities — 103
- 7–1 Simple Measurements of Length, English and Metric Units — 104
- 7–2 The Vernier Scale on Length Measuring Instruments — 108
- 7–3 Simple Measurements of Mass, English and Metric Units — 112
- 7–4 Measurement of Time Intervals — 120
- 7–5 The Calibration of Measuring Instruments — 123
- 7–6 Absolute and Relative Uncertainties in Measurements — 128

Chapter 8
Experimental Data — 132
- 8–1 Fundamental and Derived Quantities — 133
- 8–2 The Experimental Method, Control, and Experimental Variables — 134
- 8–3 Recording Experimental Data — 137
- 8–4 The Mean of a Set of Measurements — 141
- 8–5 The Mean Deviation of a Set of Measurements — 144
- 8–6 The Standard Deviation of a Set of Measurements — 148
- 8–7 Probable Error — 156

Chapter 9
Graphical Analysis of Experimental Data — 159
- 9–1 Coordinate Axes — 160

9–2	Graphs of Points; the Smooth Curve	164
9–3	Graphs of Linear Functions	169
9–4	Linear Plots of Non-linear Functions	174
9–5	Semi-logarithmic Graph Plots	179
9–6	Log-Log Graph Plots	186

Chapter 10
A Scientific Experiment   *190*

10–1	Identification of Variables	191
10–2	Design of the Experiment	193
10–3	Collection of Data	196
10–4	Data Analysis	200
10–5	Experimental Results	204
10–6	Conclusions	209

Chapter 11
Reading Technical Materials   *212*

11–1	Reading Definitions	213
11–2	Graphs, Figures, and Diagrams	216
11–3	Deductions and Derivations	219
11–4	How to Study an Assignment	223

Chapter 12
Solving Technical Problems   *226*

12–1	Reading a Problem	227
12–2	Selecting Relationships for Problem Solution	231
12–3	Deducing Needed Relationships	233
12–4	Computation in Problem Solving	236
12–5	Systematic Procedure for Problem Solution	240

Chapter 13
Sources of Technical Information   *243*

13–1	Textbooks	244
13–2	Parts of a Book	245

13-3	The Library Catalog System	248
13-4	Library Materials in Science and Technology	253

Answers to Selected Problems   *255*

Index   *265*

Quantitative Aspects of Science and Technology

1 Physical Quantities and Definitions

It is not uncommon for a word to have more than one meaning. In routine conversation, words seldom have exactly the same meaning for each person who uses them. Even such words as *energy, power, force, pressure,* and *charge* are used by people in very general ways with only vague meanings attached to them. But to a scientist or technician, words must become exact. A word like energy, work, force, or pressure must mean the same thing to every scientist or technician who uses the word. Otherwise, scientific communication, duplication, or verification would be impossible. The great advances which have been made in science and technology during the last few hundred years have been possible primarily because of the exactness of the definitions of words.

1-1 Fundamental Physical Quantities

It is a remarkable fact that the vast technology which exists today is based on a relatively small number of fundamental scientific principles. It is even more remarkable that among all these principles, each involves at most only four fundamental physical quantities: measures of *space*, *time*, *mass*, and *electric charge*. These four quantities, in their various combinations, somehow make up all the physical laws that have made our modern technology possible. All other physical definitions must be made in terms of these quantities.

What is space? This question is too difficult to answer. How can we devise a measure for space? Perhaps this task is easier. Take a straight stick of wood and cut it to some convenient length. This stick of wood can be called a unit of length, and one could give it a name, like the word "bleep". If one wanted to measure the length of a room, he would count how many "bleeps" would fit along the length of the room. It is not likely that an even number of bleeps would fit, therefore he would need to subdivide the bleep into smaller units. Perhaps he would first mark the stick off into ten equal parts, from which he could then make smaller sticks. The prefix *deci* indicates one tenth, thus each of these subdivisions could be called a "decibleep".

When one measures the length of the room, he may count 12 bleeps plus 4 decibleeps, and still find a little space left to the end of the room which is smaller than a decibleep. It is clear that he must subdivide the decibleep. Assume that he places ten equally spaced marks on the decibleep. This means that it would take 100 of these smaller marks to make one bleep. Just as it takes 100 cents to make one dollar, the prefix *centi* could be used to say that it takes 100 "centibleeps" to make one bleep. Now when the length of the room is measured, he finds that it is 12 bleeps, 4 decibleeps, and 3 centibleeps, but with a very small space still left, too small to fit in another centibleep, but large enough to observe.

One can see that another subdivision must be made, perhaps dividing the centibleep into ten more parts. Because there would then be 1000 of these smallest divisions in one bleep, we use the prefix *milli*, one one-thousandth, and call each of the smallest divisions a "millibleep". Now the room measurement gives 12 bleeps, 4 decibleeps, 3 centibleeps, and 7 millibleeps. It is possible that even using the millibleep, one can see a very minute distance not yet accounted for. One could keep subdividing a measuring instrument until the divisions could no longer be seen, even with a microscope. He must decide how precise his instrument must be, and quit subdividing at that point.

Assume that the measurement of 12 bleeps, 4 decibleeps, 3 centibleeps, and 7 millibleeps is sufficiently precise. Because each of the subunits is 1/10th the next larger unit, we can write this measurement in our base 10 number system. It would be simply 12.437 bleeps.

Having the unit of length, the bleep, one could preserve it very carefully, making copies to send to other people all over the world. If scientists reported measurements in bleeps, and merchants sold goods in terms of bleeps, it would be very important that the unit remain the same, not changing for any reason. Perhaps an exact copy should be made of some substantial metal, and this copy placed in a controlled environment where it would be secure. Then other secondary "standards" could be made and located throughout the world. One can see how a standard unit of length originates.

The international unit of length, the *meter*, is defined very much as the bleep in the preceding discussion. The selection of a unit of length is arbitrary. If everyone accepted the bleep, it would be just as good as the meter. Its use would make absolutely no difference in the fundamental laws of science.

Measures of time must also be established in an arbitrary way. Because we are usually concerned with time intervals, a standard interval must be found. Something that repeats its motion regularly must be found. Surprisingly, there exist almost no naturally repetitive phenomena on the earth. We must look to the sky, where we see the moon moving in its regularly repeating orbit, and from which we can show that the earth moves in repeating regularity about our sun. It is in terms of the earth's motion about the sun that we have established our unit of time, a unit which has been subdivided and further subdivided for our own convenience. But again, the selection of the *second* or *year* as a time unit is arbitrary. The physical laws would be completely unaffected by a change of time unit to some other repetitive phenomenon. To copy a unit of time, one must devise an instrument which repeats its motion simultaneously with the standard. In a practical way, clocks and watches provide reasonably good copies of our standard time unit, and they can be corrected by comparison with the movement of our earth about the sun.

The concept of mass is somewhat more difficult than time and space for the non-scientist to understand. He often confuses mass and weight, yet they are not the same thing. He also tends to think incorrectly of mass only as a measure of the amount of matter in something. In an elementary sense, the quantity called *gravitational mass* is closely related to the amount of matter in something, but *inertial mass*, a quantity equivalent to gravitational mass at speeds much less than the speed of light, requires an understanding of other physical concepts for its definition. For the time being, we shall consider only gravitational mass.

A unit of gravitational mass is arbitrary in the same sense as the unit of length and the unit of time. Once it is selected, it must be subdivided; then some method of comparison must be used. One might compare two masses by holding one in each hand to see how "heavy" they feel. This method is not too exact, therefore we take advantage of a balance. If a standard mass balances with another unknown mass, we say that they are equal in gravitational mass, or, more briefly, that they have the same mass. In this way it is possible to find the mass of any unknown by comparing it with standard masses using a balance.

The fourth fundamental physical quantity, electric charge, is somewhat more difficult to standardize and measure. The measurement of electric charge must be in terms of derived physical quantities not yet discussed, and its unit must be defined in terms of the three physical quantities already mentioned, but in a rather complicated way. For that reason, electric charge will not be discussed any further in this textbook.

PROBLEMS

1. Take a piece of wood of arbitrary length. Divide it into 100 equal divisions as accurately as possible. Give this unit of length a name; then measure the length and width of a room with this instrument. Divide the length by the width and obtain a number. Perform the same measurement and division of length by width using a meter stick or yardstick. What can be concluded from these results?

2. How many millibleeps are there in 12 centibleeps?

3. How many centimeters are there in 5 millimeters?

1-2 Physical Definitions and the Operational Process

Physical laws are concerned about physical quantities. To understand science, one must first know and understand the fundamental physical definitions. But in science a definition cannot be made simply in terms of words—not if it is to be useful. To make definitions exact and useful, the scientist chooses to use what he calls the *operational process*. To define a quantity by the operational process, one must simply describe how the quantity is to be measured. If it cannot be measured, it either does not exist, or it can have no real importance in science.

As an example, the quantity *average speed* can be defined by the operational process in the following way. To find the average speed of an object, one must measure the distance traveled along some path, and divide it by the time required to travel through that distance. Another example would be the quantity *pressure;* but this definition brings up a problem. *Pressure* is defined as the force acting on a surface divided by the area of that surface. So far, neither of these quantities, force or area, has been defined. Thus pressure is a quantity which is defined in terms of other quantities which must first be defined. This sort of thing happens throughout science. Many definitions are made in terms of the fundamental quantities space, time, and mass; but other definitions are then made in terms of definitions already made, so that the process gets rather complicated.

The basic point remains. To define a physical quantity, describe how it is to be measured in terms of other previously defined physical quantities.

1-3 Mathematics in Science and Engineering

Measurement of physical quantities is inherent in the definition of those quantities. The concept of measurement, and number, points immediately to the requirement that science be mathematical. Physical definitions are stated mathematically. For example, average speed, defined above in words, would be defined mathematically by the equation

$$\bar{v} = d/t$$

Pressure would be defined by the equation

$$P = F/A$$

Each symbol represents a physical quantity, and the slant bar means to divide. These equations could be stated in words, but as such definitions become more complicated, the use of words becomes impossible.

The process of logical deduction is used to a great extent in science. To use this process effectively, definitions must be exact, and therefore mathematical, requiring that the logical process itself be mathematical. To understand technology and science, one must understand mathematics.

1-4 Units for Fundamental Physical Quantities

The fundamental unit of space measure with which Americans are most familiar is the *foot*. It is subdivided into inches and parts of an inch. This English system is based on the rather unwieldly divisions of eighths, sixteenths, thirty-seconds, and sixty-fourths, which are not at all compatible with our number system based on tenths, hundredths, and thousandths. It is rather easy to see why the English system was developed, for it is considerably easier to divide something continually in half than to divide it into tenths. But there can be no question that a division of the unit of length into multiples of 10 would make measurement much easier. For this reason the scientist uses the metric system as his system of measurement.

The basic unit of length in the metric system is the *meter*. It is defined as the distance between two lines ruled on a meter bar of platinum-iridium metal kept at the International Bureau of Weights and Measures in Sèvres, near Paris, France. An accurate copy of this international standard is kept at the National Bureau of Standards in Washington D. C. Even the common inch is now defined in terms of the meter; specifically, 1 inch is defined as exactly 0.0254 meters.

Subdivisions of the meter are in units of ten. Thus the meter is divided into 100 equal distances called centimeters. Each centimeter is divided into 10 equal distances called millimeters. Meter sticks do not usually show any finer marks than millimeters, but further subdivision is possible. A meter is only slightly longer than a yard. A centimeter is about as wide as a finger, and a millimeter is about as wide as a pencil lead. As a person uses the metric scale in measurement, he gains a feeling for the sizes of the unit and subunits, just as he did for the English system in common usage.

As has already been indicated, the concept of mass is not simple. The two aspects of mass include the ideas of inertia and the amount of matter in something. We shall consider here the quantity gravitational mass, which measures roughly the amount of matter in something. Just as there is a standard meter bar in France, there is also a standard unit of gravitational mass, the *kilogram*. The standard kilogram is a platinum-iridium metal cylinder, kept at the International Bureau of Weights and Measures at Sèvres, France. A replica of the kilogram is also kept at the Washington D. C. Bureau of Standards. The kilogram is about 2.205 pounds in weight. One one-thousandth of a kilogram is called a gram, and one one-millionth of a gram is called a microgram. A pencil eraser has a mass of about one gram; a microgram would be so small that one could not feel the weight. It is important to remember that mass and weight are not the same thing. To compare masses, one must use a balance.

Time intervals are measured in terms of the motion of the earth about the sun. Time units are defined in terms of a *solar day*, a day determined by the repetition of the position of the sun with respect to some point on the earth. Using careful astronomical data, one year has been defined as 365.24219879 days. The time unit called second is defined as 1/86400 of a solar day. The minute, of course, is 60 seconds and the hour is 60 minutes. These units of time are precisely the same as those in ordinary use.

The system of units described above is the beginning of what is called the MKS system. The M stands for meters, K for kilograms, and S for seconds. As one studies science, he combines these units to make other more complicated units which belong to that system. A knowledge of the fundamental units is essential to understanding physical science. Prefixes for units in terms of base-ten numbers make possible the specification of various sub-units as needed. You should memorize the basic prefixes, and practice using them with various units. Table 1-1 lists commonly-used units and prefixes.

TABLE 1-1

Unit	Abbreviation	Multiple	
Meter	m	1	m
Centimeter	cm	1/100	m
Millimeter	mm	1/1000	m
Micron	μ	1/1,000,000	m
Millimicron	mμ	1/1,000,000,000	m
Kilometer	km	1000	m
Second	sec	1	sec
Millisecond	msec	1/1000	sec
Microsecond	μsec	1/1,000,000	sec
Minute	min	60	sec
Hour	hr	3600	sec
Kilogram	kg	1000	g
Gram	g	1	g
Milligram	mg	1/1000	g
Microgram	μg	1/1,000,000	g

There are, in addition to the prefixes already mentioned, a number of other prefixes with which we should be familiar. Table 1-2 lists most numerical prefixes used in science and technology today.

TABLE 1-2

Prefix	Symbol	Multiple	
Kilo	k	1000	$=10^3$
Mega	M	1,000,000	$=10^6$
Giga	G	1,000,000,000	$=10^9$
Tera	T	1,000,000,000,000	$=10^{12}$
Deci	d	0.1	$=10^{-1}$
Centi	c	0.01	$=10^{-2}$
Milli	m	0.001	$=10^{-3}$
Micro	μ	0.000001	$=10^{-6}$
Nano	n	0.000000001	$=10^{-9}$
Pico	p	0.000000000001	$=10^{-12}$

PROBLEMS

1. How much money is 12 gigadollars?
2. How long is 12 nanoseconds?
3. China has a population of approximately 600,000,000 people. How many is this in terms of terapeople?
4. Our recent defense budget was about 60,000,000,000 dollars. How much is this in kilodollars?
5. How many nanoseconds are there in 5 milliseconds?
6. How far is one mile (5280 feet) in terms of centimeters?
7. How many microseconds are there in one year?
8. How many kilograms is 6 picograms?
9. Measure the width of this page and express your answer in terms of a. meters, b. kilometers, c. centimeters, d. millimeters, e. microns, f. picometers, and g. terameters.

2 Numerical Symbols and Operations

Science and technology have been successful because of their quantitative nature. Because a scientist or engineer not only can describe what is going to happen under a given set of circumstances, but also can tell how much it will happen, and where and when, he has a remarkable ability to control future events. He knows that when he builds a complicated electronic device, it will in fact do what he wants it to do. A television set works, and it works well most of the time. Considering its complexity and dependence upon such a large number of essential quantitative conditions, the simple fact that a television set works is remarkable. This is but one of a vast number of modern technological examples. Anything that can be described, can today be designed and built. The major reason for this capability is the quantitative nature of science and technology.

2-1 Natural Numbers

To be quantitative, one must work with numerical symbols. A *symbol* is a word or mark of some kind that people use to represent physical things. For example, the word "tree" represents a physical thing that we see outside which grows rather large and has leaves on it. But the word "power" does not represent some physical thing; it would be more complicated to explain. Everyone has at one time or another counted objects. Each of us uses symbols to represent certain numbers of things. Actually, however, we rarely are able to conceptualize very large numbers, that is, larger than ten or twenty. Even though we can symbolize the number six million by 6,000,000, none of us can actually conceive of what this symbol represents.

Our first experience with numbers is in counting. Usually we learn to count before we learn written symbols to represent those quantities. We say the words "one", "two", "three", and so forth. Each of these words is a symbol representing a particular quantity. When we learn to write the symbols "1", "2", "3", and so forth, we are writing what are called *numerals*. The numerals represent numbers. But nobody can write a number, for a number is the quantity we want to represent.

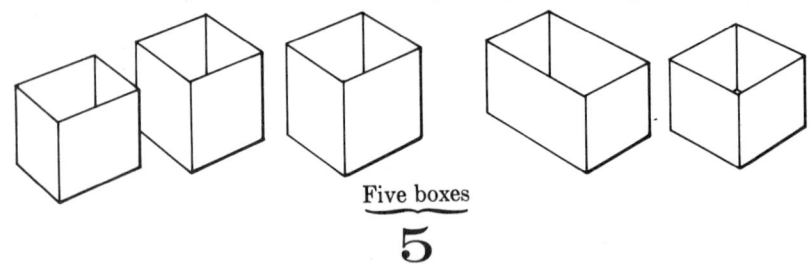

Five boxes
5

Figure 2-1. *A numeral is a symbol which represents a quantity called number.*

The numbers "one", "two", "three", etc., are called *natural numbers*. They are the ones we use in counting things. We represent these numbers by the symbols called numerals, "1", "2", "3", etc. In this book, we will use the numerical symbols without quotation marks, assuming that those symbols stand for numbers. We will also use the name "number" both for actual numerals, and for the quantities they represent. Thus we have a set of numerals which starts at "1" and increases by "1" indefinitely:

$$1 \quad 1 + 1 = 2 \quad 2 + 1 = 3 \quad 3 + 1 = 4$$

[§ 2-1] Natural Numbers 11

and so on. What these numerals, 1, 2, 3, 4, 5, 6, 7, 8, etc., represent are called *natural numbers*.

Once aware of natural numbers, people began to do things with the symbols that represented them. Sometimes two sets of things would be counted separately; then it was desired to combine the two sets and know how many there were without recounting the larger set. It soon became possible to perform this operation. It has been called *addition*. Every college student knows how to add natural numbers, and he knows that the symbol which represents this process is the plus sign, +. If he sees 5 + 8, he has learned by memorization, without having to count the two sets, that the sum is 13. He also has learned to add such numerals as 34 + 28, because he has been taught certain rules of addition that work. Yet he may not understand why they work. A detailed study of number theory would not be appropriate here, therefore it will be assumed that the reader can perform the operation of addition of natural numbers.

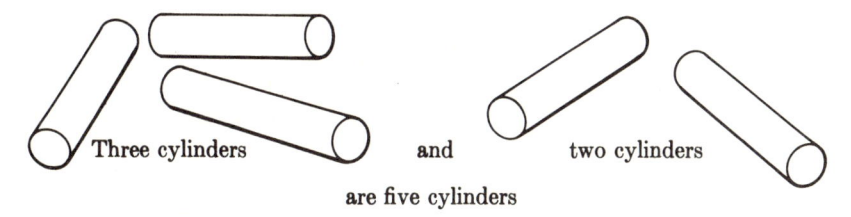

Three cylinders and two cylinders
are five cylinders

Figure 2-2. *Addition of natural numbers consists of combining two smaller sets to form a larger set. Counting the larger set gives the sum of the numbers representing the smaller sets.*

Another process, somewhat less understood by some, is that of subtraction. They think that they understand subtraction because they can successfully subtract some kinds of numerals, when a smaller numeral is taken from a larger one. But when a larger numeral is to be taken from a smaller numeral, some people are confused. It seems to them that one should not be able to take away more than there is. This confusion arises because of a basic misunderstanding of the subtraction operation, and of our number system.

The mathematical definition of subtraction is made in terms of addition. It is not a separate process. The mathematician would define subtraction in the following way. If the sum of two numbers has the same value as a third number, the first number is said to be the *difference* of the second and third ones mentioned. This definition sounds rather confusing. Let us use symbols to say the same thing. Let X represent the first number, Y represent the second number, and Z represent the third number. If $X + Y = Z$ (or $Y + X = Z$), then X is called the difference of Y and Z. As long as Z is larger than Y, this definition

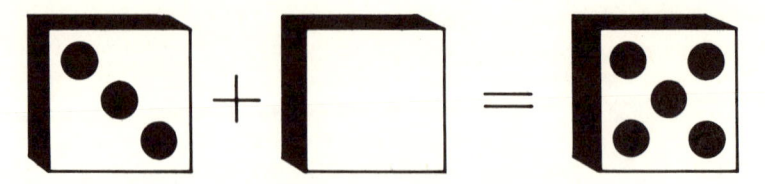

Figure 2-3. *The steps taken to "fill in" the missing dots on the center die face define the process of subtraction.*

permits us to perform the familiar process of subtraction. But if Y is larger than Z, familiar ideas of subtraction are inadequate. What is bothersome is that such subtractions give numerals which no longer represent natural numbers. For example, because $6 + 2 = 8$, 6 must be the difference of 8 and 2, or $6 = 8 - 2$. Here we have used the negative sign to mean subtraction. Using that sign, we would say that if $X + Y = Z$, then $X = Y - Z$ is a definition of subtraction. If the difference $9 - 5$ is to be found, the definition requires that some number be found which, when added to 5, gives the number 9, $5 + 4 = 9$. But finding the difference $3 - 8$ seems strange, because the definition requires that we find a number which, when added to 8, gives 3. No natural number can do this. As we will see in the next section, the difference $3 - 8$ is an integer.

PROBLEMS

1. Change the following statements from mathematical to verbal (word) form using the definition of subtraction. For example, $3 - 2 =$ is read, "the number to be added to 2 to give 3 is".
a. $5 - 3 =$ b. $8 - 7 =$ c. $4 - 12 =$ d. $X - 4 =$
e. $3 - X =$ f. $X - Y =$

2. Using only *natural numbers*, perform the following indicated operations. If there is no result in the set of natural numbers, indicate the problem by the statement "no solution".
a. $3 - 50$ b. $9 - 3$ c. $6 + 129$ d. $8/4$ e. $4/5$ f. $4 - 7$
g. $1024/4$ h. $(70732)(2883432)$

2-2 Integers

Numbers other than natural numbers are needed in science and technology. Everyone is familiar with the idea of negative temperatures, like −34 degrees. Notice that this is a *new* way of using the negative sign. It does not mean what it does when used to indicate subtraction. In a similar way, any quantity which is below, or smaller than, some arbitrary zero level is called negative. If you decide that ground level is at a certain height, four feet below that height is called −4, seven feet below it is called −7, etc. Such numerals are also useful to represent direction. If it is decided that a certain direction is positive, the opposite direction can be negative. Thus if a car is moving at 50 mph due north, a car moving 50 mph due south could have a velocity of −50. The negative sign has become an extremely useful symbol to represent such quantities. Fortunately, the mathematician had developed the operations dealing with such numbers long before they were needed by science and technology.

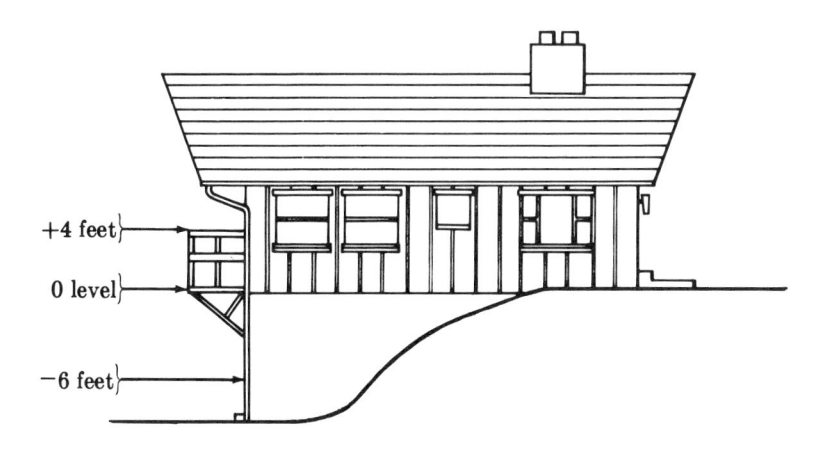

Figure 2-4. *Quantities below zero are called negative.*

The mathematician found that certain problems existed for which there were no solutions in the set of natural numbers. He could write an equation like $2 + X = 0$, but there was no natural number for X which would make this algebraic statement true. Thus he invented a new set of numbers, each of which could be added to a natural number to give zero. If a represents a natural number, $-a$ is *defined* as the unique number which, when added to a, gives zero. Using these symbols in an algebraic statement, the definition says $-a + a = 0$. For example, -2 is defined as the number to be added to 2 to give zero, $-2 + 2 = 0$. The mathematician has given this kind of number a fancy name, the *additive inverse*. Additive inverse is hard to pronounce, but it merely describes a number to be added to another number to give zero. In that sense, -2 is the additive inverse of 2, and vice versa. The number 0 has also been given a fancy name, the *additive identity*. The reason is simple: for any number N, $N + 0 = N$. The operation of adding zero to the number N does not change N. It "retains its identity". With the invention of the negatives of each of the natural numbers, the new set of numbers, "... $-8, -7, -6, -5, -4, -3, -2, -1, 0, +1, +2, +3, +4, +5, +6, +7, +8, \ldots$", has been called *integers*. An *integral* number is simply one of these numbers.

PROBLEMS

1. Using only *integers*, perform the following indicated operations. If there is no result in the set of integers, indicate the problem by the statement "no solution".

    a. $4 + 50$ b. $4/5$ c. $+8/2$ d. $+3/7$ e. $+1024/2$

2. Find the additive inverse of each of the following numbers.

    a. 4   b. -8 c. 12.32   d. π

2-3 *Rational Numbers*

Numbers other than integers also occur in our experience with measurements. The temperature may not be exactly $-32°$; it might be between $-32°$ and $-33°$. The distance of a point above ground level may not be 6 feet; it might be somewhere between 6 and 7 feet. An electrical measuring device, like a voltmeter, indicates readings continuously between two numbers like 4 and 5 or -7 and -8. Those in-between numbers, which we write as fractions or decimals, without thinking much about their meaning, were also invented long ago by the mathematician. He was able to define them using the numbers he already had, the integers.

People had long been accustomed to dividing natural numbers, as long as the dividend was larger than the divisor. The number 12 divided by 3 meant to divide 12 objects into 3 sets of 4 each. A person could understand this division in terms of counting real objects. But the

[§ 2-3]   Rational Numbers   15

notion of dividing 5 by 12 would seem absurd. After all, how could one divide only 5 objects into 12 sets. As long as one thought in terms of single, unbreakable objects, this idea was strange. But the process of division needed to be expanded, and there are some things which could be broken into parts; thus division of this sort should make sense.

The mathematician had a further problem in solving equations like $2X = 1$, $3X = 1$, $-4X = 1$, etc. There was no number X in the set of integers for which any of these equations would be true. Thus he invented numbers analogous to the negatives in addition. He needed a number which, when multiplied by an integer, would give the number one. He invented these unique numbers, which he symbolized with a fraction bar as in $1/2$, $1/3$, $-1/4$, etc. They were called *reciprocals* of the numbers in the denominators. In this way, $(2)(1/2) = 1$ was the definition of the reciprocal $1/2$. He also gave the reciprocal a fancy, almost unpronounceable, name, the *multiplicative inverse*. And of course, to add the same mystery to the number one, it was called the *multiplicative identity*. Since, for any number N, $(N)(1) = N$, the multiplication of the number N by 1 leaves N unaffected; it "retains its identity".

Other problems existed for the mathematician. He could not solve equations like $3X = 2$, $-4X = 7$, $5X = -2$, etc. But with the invention of reciprocals, he could solve these equations. All he had to do was multiply both sides of each equation by the reciprocal of each multiple of X. For example, in the equation $3X = 2$, multiplying each side by the reciprocal of 3, $1/3$, he had $(3)(1/3)X = (2)(1/3)$. But by definition of the reciprocal, $(3)(1/3) = 1$. Thus he had $X = (2)(1/3)$. If there were some way of multiplying 2 by $1/3$, the problem would be solved. To solve this problem, he invented an entirely new set of numbers. He had already decided that zero times any number would be zero, therefore zero could have no reciprocal. That is, $(0)(1/0)$ could not equal 1, thus $1/0$ would be meaningless, undefined. But for every other integer N, there would be a reciprocal $1/N$. He found ways of expressing these reciprocals as decimals, and we have already learned rules for determining them as decimals. Rules were also determined for multiplying one integer by the reciprocal of another integer. These products then formed what is called the set of *rational numbers*. Let us define a fraction A/B by the product $(A)(1/B)$; then as a formula definition we have

$$A/B = (A)(1/B) \qquad (2\text{-}1)$$

We usually write an integer multiplied by a reciprocal as a fraction like $-3/7$, but it must be remembered that this fraction means nothing more than $(-3)(1/7)$. The rational numbers, expressed as fractions, are numbers like...$1/3$, $-2/7$, $-3/1$, $5/7$, $9/4$,..., where any integer is "over" any other integer (except zero). Expressed as decimals, the rational numbers are "in-between" the integers, although the integers are also rational numbers, since $-3/1 = -3$, $5/1 = 5$, $77/1 = 77$, etc.

We have seen how the mathematician began with the set of natural numbers, and extended it to include the negatives of the natural numbers and zero to form the set of integers. Then he invented the reciprocals of integers (except zero) so that he could construct rational numbers. Because the set of rational numbers contains all integers and, of course, all natural numbers, any rules of operation for addition, multiplication, subtraction, or division applying to the set of rational numbers will also apply to any of the subsets like integers or natural numbers. But the rules of operation learned in elementary school for natural numbers or positive rational numbers cannot be applied to all the rational numbers. This is precisely why students have difficulty understanding algebra. They try to apply simple rules that they learned for positive integers or positive rational numbers to other rational numbers where the rules will not work.

PROBLEMS

1. Using only *rational numbers*, perform the following indicated operations. If there is no result in the set of rational numbers, indicate the problem by the statement "no solution".

    a. $+3/4 + 2$ b. $4/5$ c. $5/\pi$

2. Find the multiplicative inverse of each of the following numbers.

    a. 5   b. 1/5   c. 0.3   d. $-7/2$

2-4 Addition of Rational Numbers

A person first learns to add natural numbers. He begins by counting. If he is to add 4 and 5, he places these two sets together and counts them to get 9. As long as the sum is less than 10, his fingers serve nicely as objects to be counted. Eventually he memorizes the correct sums of pairs of the numbers 0, 1, 2, 3, 4, 5, 6, 7, 8, 9. He will also have learned to write numbers like 58 and 13, or 20,034 using only the ten digits and place holders.

Although seldom understanding the basis of our number system, we add numbers like these quite successfully. We have learned, by drill and rote memory, certain rules that work. The extension of these rules to all positive rational numbers is relatively simple, and we soon learn to handle these numbers as well. We need only learn certain ways of

[§ 2-4] Addition of Rational Numbers 17

working with decimals. Since addition is based on counting, what we have learned can be of great help to us in algebraic addition, but we must realize that our experience with addition has been with a restricted set of numbers and that the rules are more complicated for application to all the rational numbers. To arrive at correct rules of addition for all rational numbers, we shall examine certain specific examples.

Consider the various kinds of rational numbers that might be added. There are four cases to examine. In Case I, both numbers are positive. In this case, the *familiar process* of addition of positive rational numbers is used, that of simply counting. For example, (+7.3) + (+6.8) is found by adding the tenths to get eleven of them, and the units to get thirteen of them. Since eleven tenths is ten tenths, or one, plus one tenth, the answer must be 13 + 1 plus 0.1, or 14.1. This is the familiar addition process that almost everyone can perform without even thinking about it.

In Case II, the numeral with the larger absolute value is negative, the other being positive. As a specific example, we could have (−9.7) + (+4.2). To determine a rule for adding numbers like these, we make the equation

$$(-9.7) + (+4.2) = N$$

where N is the sum of these two numbers. From the definition of the equal sign, we can add, by familiar (positive rational number) addition, any quantity to both sides of this equation. We know by definition of negatives that −9.7 means (−9.7) + (+9.7) = 0. Therefore, we can "get rid of" the −9.7 in the equation above by adding (+9.7) to both sides of the equation. Doing this addition, we have

$$\underline{(-9.7) + (+9.7)} + (4.2) = N + (+9.7)$$

But, by definition of negatives, the underlined quantity is zero. Thus,

$$(+4.2) = N + (+9.7)$$

If we then subtract 4.2 from both sides of this equation, we have

$$4.2 - 4.2 = N + 9.7 - 4.2 \quad \text{or} \quad 0 = N + 5.5$$

Notice that in the addition and subtraction operations we have used the familiar processes, since they apply quite well as used here with positive rational numbers. The result, $0 = N + 5.5$, is true only for $N = -5.5$ by the definition of negative numbers.

In Case III, the number having the smaller absolute value is negative and the larger number is positive. As an example, we could consider the sum

$$(+8.8) + (-4.5) = N$$

Adding (+4.5) to both sides of this equation, we have

$$(+8.8) + \underline{(-4.5) + (4.5)} = N + 4.5$$

But, again, by definition of the negative number −4.5, the underlined quantity is zero. The equation then reduces to

$$(+8.8) = N + 4.5$$

Substracting 4.5 from both sides, the result is

$$8.8 - 4.5 = N + 4.5 - 4.5 \quad \text{or} \quad 4.3 = N$$

Case IV occurs when the signs of both numbers are negative. As an example, consider the sum

$$(-5.2) + (-3.1) = N$$

Adding (+5.2) and (+3.1) to both sides of this equation gives

$$\underline{(-5.2) + (+5.2)} + \underline{(-3.1) + (+3.1)} = N + 5.2 + 3.1$$

But the underlined quantities are each zero by the definition of negative numbers. Therefore, $0 = N + 8.3$. By the same definition of negative numbers, N must be −8.3.

Look at the results of the four examples considered above.

$$\begin{aligned} &\text{I} \quad (+7.3) + (+6.8) = (+14.1) \\ &\text{II} \quad (-9.7) + (+4.2) = (-5.5) \\ &\text{III} \quad (+8.8) + (-4.5) = (+4.3) \\ &\text{IV} \quad (-5.2) + (-3.1) = (-8.3) \end{aligned}$$

From these examples, we can devise addition rules for rational numbers. From examples I and IV it is seen that when the signs are alike, we simply add the numerical absolute values, giving the result the sign common to the numbers involved. In examples II and III, where the signs are unlike, the numbers are *added* by finding the difference of the numerical absolute values, giving the result the sign of the larger of the two. These rules can be used when adding any two rational numbers. We must realize that this addition process is different from what we are ordinarily accustomed to using.

PROBLEMS

1. Using the same steps followed in Section 2-4, *prove* that

$$(-7.7) + (5.1) = (-2.6)$$

2-5 Subtraction of Rational Numbers

The definition of subtraction was made in Section 2-1. It was stated that if $X + Y = Z$, then X is called the difference of Y and Z, or, $X = Z - Y$. From this definition, we can construct a rule for subtraction of rational numbers.

Let us again consider specific examples. In the first, let

$$(+6.3) - (+4.2) = N$$

Of course, this is the familiar positive rational number subtraction problem for which N is simply 2.1. But consider the example

$$(-8.5) - (+5.2) = N$$

In this case, the definition above requires that

$$N + 5.2 = -8.5$$

Adding (+8.5) to both sides of this equation, and using the definition of the negative number -8.5, we have

$$N + 5.2 + 8.5 = (-8.5) + (+8.5) = 0$$

From this equation $N + 13.7 = 0$, and, again, by the definition of the negative number (-13.7), $N = -13.7$, and the result is

$$(-8.5) - (+5.2) = -13.7$$

As another example, let

$$(+7.7) - (-4.1) = N$$

In this case, using the definition of subtraction gives

$$(+7.7) = (-4.1) + N$$

Then, adding (+4.1) to both sides of this equation, there results

$$(+7.7) + (+4.1) = \underline{(-4.1) + (+4.1)} + N$$

Again, the definition of negatives requires that the underlined quantity be zero, so that the equation becomes $N = 11.8$, or, finally,

$$(+7.7) - (-4.1) = +11.8$$

Now let us examine the results of these examples.

$$(+6.3) - \underline{(+4.2)} = (+2.1)$$
$$(-8.5) - \underline{(+5.2)} = (-13.7)$$
$$(+7.7) - \underline{(-4.1)} = (+11.8)$$

Notice that the result in each case is the same as one would get by changing the sign of the subtrahend (underlined) in each case and performing the process of addition using the rules previously given. The additions would appear as follows:

$$(+6.3) + (-4.2) = (+2.1)$$
$$(-8.5) + (-5.2) = (-13.7)$$
$$(+7.7) + (+4.1) = (+11.8)$$

The rule for subtracting rational numbers can now be stated. If N and M are rational numbers, $N - M$ is found by changing the sign of the subtrahend M and adding, using the rules of addition already developed.

PROBLEMS

1. Using the same steps followed in Section 2-5, *prove* that

$$(+8.8) - (-5.3) = (+14.1)$$

2-6 *Multiplication of Rational Numbers*

The process of multiplication of rational numbers must lead to special rules, just as did addition and subtraction. For positive rational numbers, multiplication is performed just as is commonly done, but difficulties arise when one or both of the two factors are negative.

Consider the case when one factor is positive, and the other negative. Using the example $(+6)(-5) = N$, we wish to find the product N. The definition of (-5) means $(+5) + (-5) = 0$. Multiplying both sides of this equation by $(+6)$, one has

$$(+6)[(+5) + (-5)] = (6)(0) \quad \text{or} \quad (+30) + (6)(-5) = 0$$

But we already know, from the definition of the unique number (-30), that $(+30) + (-30) = 0$. It must then be true that

$$(+6)(-5) = (-30) = N$$

From this example it can be seen that for products of factors with unlike signs, the result is negative, the numerical value of the result being the ordinary product of the absolute values of the two factors.

In the case when both factors are negative, look at the example $(-4)(-3) = N$. Again, (-4) means $(-4) + (+4) = 0$. Multiplying both sides of this equation by (-3), it becomes

$$(-3)[(-4) + (+4)] = (-3)(0) \text{ or } (-3)(-4) + (-3)(+4) = 0$$

But, by the rule we just developed for products of unlike signs, the second pair of factors above is equal to -12. That equation then is

$$(-3)(-4) + (-12) = 0$$

Then by the definition and uniqueness of (-12), $(+12) + (-12) = 0$, we must have

$$(-3)(-4) = (+12) = N$$

From this example, when the signs are both negative, the product is positive. This was also the result when both sides were positive.

Using the examples stated above, rules for multiplication of rational numbers can be described. In any case, the numerical value of the product is the same as found in ordinary multiplication—the product of the absolute values of the two factors. If the signs are alike, the product is positive. If the signs are unlike, the product is negative.

PROBLEMS

1. Using the same steps followed in Section 2-6, *prove* that

$$(-7)(+6) = (-42)$$

2-7 *Division of Rational Numbers*

The process of division of rational numbers is the inverse of multiplication. Division is defined in the following way. To divide a number N by a number M, multiply N by the reciprocal of M. Written with symbols, the definition says $N/M = (N)(1/M)$. The fraction bar, /, instead of the symbol ÷, is used to denote division. For example, 5 divided by 7 would be written $5/7$, or $(5)(1/7)$.

Because division is just a form of multiplication, the rules for signs must be the same as in multiplication. When dividing two rational numbers having unlike signs, the result is negative. When the signs are alike, either divisor and dividend both positive or both negative, the result is positive. The process of numerical division of rational numbers is carried out with the techniques we have learned in elementary school, using decimals as needed.

As an example of division, $-3/5$ is found by computing $(-3)(1/5)$. But $1/5 = 0.2$, therefore

$$(-3)(0.2) = -0.6$$

We have divided (-3) by (5) to get $-3/5 = -0.6$.

2-8 Summary of Rules of Operation on Rational Numbers

Although no attempt has been made here to prove rigorously each rule stated, the examples have been developed in a logical way. To understand those examples by this heuristic method requires careful and thoughtful study. If not clearly understood, those examples should be reexamined more carefully.

The resulting rules of operation for rational numbers are as follows.
Addition
   Signs Alike. Sum the two absolute values; use common sign.
   Signs Unlike. Find the difference of the two absolute values; use sign of the larger absolute value.
Subtraction, $N-M$
   Change the sign of the subtrahend M, then perform *addition* using the rules stated above.
Multiplication
   Signs Alike. Find the product of absolute values of factors; the sign is positive.
   Signs Unlike. Find the product of absolute values of factors; the sign is negative.
Division, N/M
   Find the reciprocal of the divisor M, then perform *multiplication* using the rules stated above.

PROBLEMS

1. Carry out the following indicated operations.
   a. $(+5.2) + (-8.7)$ b. $(+7.9) + (-3.1)$ c. $(+8.8) - (+4.4)$
   d. $(+9.2) - (-4.4)$ e. $(+8.4) - (+9.9)$ f. $(-3.3) - (-6.6)$
2. Carry out the following indicated operations.
   a. $(+5)(-3)$ b. $(-6)(-7)$ c. $(-3.1)(-4.2)$ d. $+7/(-2.1)$
   e. $(-8)/(-2)$ f. $-448/4$

2-9 Irrational Numbers

Rational numbers are the kinds of numbers which our decimal system is capable of representing. All computations involve these rational numbers. But there are other kinds of numerical quantities that exist, for which we need other symbols. These numerical quantities, which cannot be stated exactly as rational numbers, are called *irrational numbers*. Fortunately, we can approximate irrational numbers with rational numbers to whatever degree of accuracy we may need.

The circumference of a circle divided by the diameter of that circle gives a quantity which is irrational. Since we cannot represent it with a rational number (a decimal), we symbolize it with the symbol pi, π. When we must use it in computation, we select a rational number that is close to π, as is 3.1415926. This number is rational, and it is not π. But it is close to π.

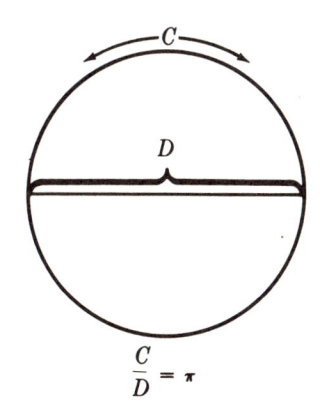

Figure 2-5.   *The ratio of C to D is π.*

When the length of the diagonal of a square is divided by the length of an edge, the result is an irrational number symbolized by $\sqrt{2}$. No one can write $\sqrt{2}$ as a rational number, but the rational number 1.414 is close to $\sqrt{2}$, so we use 1.414 when we need $\sqrt{2}$ in some computation.

There are many irrational numbers for which rational approximations must be made. The base of the natural logarithms, called e, is irrational. Many roots are irrational, for example $\sqrt{3}$, $\sqrt[3]{7}$, etc. There are an infinite number of such irrational numbers.

When all the irrational numbers are considered together with the set of rational numbers, the result is called the set of *real numbers*. These are the kinds of numbers which are used in all numerical compu-

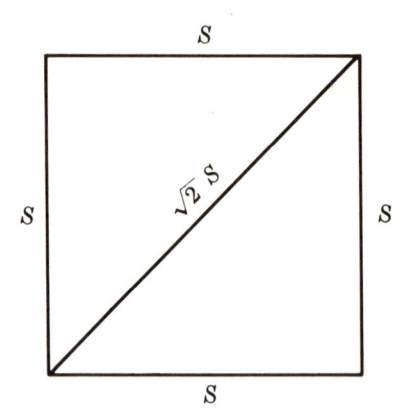

Figure 2-6. *The diagonal of a square is $\sqrt{2}$ times one edge.*

tations. There do exist other kinds of numbers, like the solution to the equation $X^2 + 2 = 0$. This more general set, called the set of *complex numbers*, includes the real numbers. However, we will not examine complex numbers in this book.

PROBLEMS

1. Carefully construct 3 different sized circles on paper. Measure the diameter of each one. Then, with a piece of string, measure the circumference of each one as best you can. Divide the circumference by the diameter for each circle. What does this rational number approximate?

2. Carefully construct a square figure of any size on paper; then measure the diagonal and an edge. Divide the diagonal length by the edge length. What does this number approximate?

3 Computation with Power of Ten Notation

One of the major difficulties encountered when studying a first course in physical science is the simple matter of computation. When investigating natural laws, the size of numbers encountered ranges from extremely small to enormously large. Yet the average person seldom encounters either very large or very small numbers. As a consequence, he never learns how to handle such numbers, how to multiply or divide them, or even how to express them. However, such an ability is essential for one to begin a serious study of technology or science.

3-1 Powers of Ten

How can one express a large number in an easier way than writing it out? The distance to the nearest star is about 10,000,000,000,000,000 m. One can imagine the difficulty of describing distances to the more remote stars by writing the quantity in this ordinary way.

To write such large numbers more easily, it is necessary to recall a few fundamentals of elementary algebra concerning exponents. But to talk about algebraic quantities, certain definitions must be made.

The following definitions are essential to an understanding of exponents.

Positive—A number is positive if it is greater than zero.

Integer—An integer is any number generated by the successive addition or subtraction of the number one to itself. For example, 6, 7, 12, and -3 are integers.

Factors—A factor is a multiple; it is a number which is to be multiplied by any other number placed next to it. If no other number is present, the factor is multiplied by the number one.

Positive integral exponent—A positive integral exponent is a positive integer placed in a position to the upper right of another number called a *base*. This exponent indicates how many factors of the base are needed to give some required number. For example, 5^3 means three factors of 5, or $5 \times 5 \times 5 = 125$.

Negative—A negative number is a number less than zero.

Consider the base 10 with positive integral exponents. How much is 10^1? The exponent indicates one factor of ten, but the one factor of ten would stand by itself, and therefore would be simply 10; $10^1 = 10$.

What about the number 10^2? The exponent requires that we show two factors of ten. Thus $10^2 = 10 \times 10 = 100$, and it is found that $10^2 = 100$. How much is 10^3? Three factors of 10 gives $10 \times 10 \times 10 = 1000$.

Because we often use the letter x to stand for something unknown in algebra, the process of multiplication is usually indicated in a way that does not use x. To show multiplication, two numbers can be placed next to each other, with parentheses around each number. For example, $(12)(8)$ means to multiply 12 by 8. Thus $(12)(8) = 96$. Sometimes a dot is placed between two numbers to mean multiplication, like $4 \cdot 6$, with the dot half way up from the bottom of each number. But using a dot might cause confusion when decimals are present. For example, what would $4.2 \cdot 3.1$ mean? You can see that there are too many dots. It would be better to write this product as $(4.2)(3.1) = 13.02$. To avoid confusion in this textbook, products of numbers will almost always be written with parentheses.

Using parentheses to indicate multiplication, how would one write 10^4? The exponent indicates that 4 factors of 10 are needed. Thus

$$10^4 = (10)(10)(10)(10) = 10,000$$

A list of the numbers considered so far reveals an interesting pattern. Look at the list carefully.

$$10^1 = 10$$
$$10^2 = 100$$
$$10^3 = 1000$$
$$10^4 = 10000$$

When the exponent is 1, there is one zero in the number on the right. When the exponent is 2, there are two zeros, when 3, three zeros, and when 4, four zeros. Without having to multiply the numbers, what should be the value of 10^7? From an examination of the pattern of exponents and zeros above, it appears that the positive integral exponent on a base of ten gives the number of zeros following the number 1. Thus $10^7 = 10,000,000$. To check this result,

$$10^7 = (10)(10)(10)(10)(10)(10)(10) \quad \text{or} \quad (100)(100)(100)(10)$$

multiplying the 10's together. But this number is

$$(10,000)(1000) = 10,000,000$$

which is what we found by the method of counting zeros. The positive integral exponent on a base of ten is the same as the number of zeros to follow 1 in the number written without exponents.

By using the above rule, determine the distance in meters to the nearest star. That distance was stated as 10,000,000,000,000,000 m. Because there are 16 zeros, it is 10^{16} m, which is easier to write with exponents than without them.

PROBLEMS

1. Using power of ten notation, express the following quantities in terms of the fundamental unit indicated.

    a. 1 km, in m   b. 1 Gm, in m   c. 1 Tsec, in sec

2. Express the following numbers in power of ten form.

    a. 10,000,000,000   b. 10   c. 1 billion   d. 1 million

3. Compare the difference between 10^3 mi and 10^4 mi with the difference between 10^{13} mi and 10^{14} mi. In each pair of quantities, one is "an order of magnitude" larger than the other. Are "order of magnitude" differences always the same amount?

3-2 Powers of Ten in Multiplication

Being able to write large numbers is convenient, but there are other aspects of this power of ten notation that make it useful to the scientist. Consider the problem of multiplying numbers. Specifically, look at the products of the following numbers:

$$(10)(10) = 100 = 10^2$$
$$(100)(100) = 10,000 = 10^4$$
$$(10)(1000) = 10,000 = 10^4$$
$$(100)(1000) = 100,000 = 10^5$$
$$(1000)(1000) = 1,000,000 = 10^6$$

But look at the factors on the left, when each is expressed using exponents:

$$(10^1)(10^1) = 10^2$$
$$(10^2)(10^2) = 10^4$$
$$(10^1)(10^3) = 10^4$$
$$(10^2)(10^3) = 10^5$$
$$(10^3)(10^3) = 10^6$$

In each case the sum of the exponents on the left is the same as the exponent on the right. This result is not an accident. It follows from the definition of a positive integral exponent. In the first product, 1 factor of 10 plus another factor of 10 makes 2 factors of 10. In the fourth product, 2 factors of 10 plus 3 more factors of 10 makes 5 factors of 10. The fact that when these numbers are multiplied the exponents are to be added follows from an algebraic rule. Symbolically, the rule says

$$(x^n)(x^m) = x^{n+m}$$

In words, it would say if two numbers *with the same base* are to be multiplied, the product is a number having the same base with an exponent which is the sum of the exponents of the two factors. By this rule,

$$(6^5)(6^8) = (6^{13})$$

But $(4^3)(3^5)$ could not be evaluated by this rule, since the bases 4 and 3 are *not* the same.

Using the above rule, what is the product of the factors $(10^6)(10^9)$? The base of 10 is the same for both factors; thus one need only add the exponents and place the result as an exponent on the common base. The answer must be

$$10^{6+9} = 10^{15}$$

By using this rule, these large numbers can be multiplied quite easily.

PROBLEMS

1. Find the products of the following factors.
   a. $(10^4)(10^7)$ b. $(10^9)(10^{23})$ c. $(10^1)(10^8)$ d. $(10^3)(10^4)$
2. Express each of the factors below in power of ten form; then carry out the indicated multiplications.
   a. $(1000)(10,000)$ b. $(100)(1,000,000)$
   c. $(10,000)(100,000,000,000,000,000)$ d. $(10)(100)$
3. Write each of the following numbers without using exponents.
   a. 10^7 b. 10^{54} c. 10^3 d. 10^1

3-3 Powers of Ten in Division

What about division of two large numbers? In mathematics, it is common practice to use the bar to indicate division instead of the symbol $\div$. For example, to show 100 divided by 10, we could show $100/10$ instead of $100 \div 10$. This convention will be used throughout this textbook.

Consider the following list of indicated divisions:

$$100/10 = \frac{(10)(10)}{10} = 10 = 10^1$$

$$1000/10 = \frac{(100)(10)}{10} = 100 = 10^2$$

$$10,000/10 = \frac{(1000)(10)}{10} = 1000 = 10^3$$

$$100,000/100 = \frac{(1000)(100)}{100} = 1000 = 10^3$$

$$1,000,000/100 = \frac{(10,000)(100)}{100} = 10,000 = 10^4$$

If nothing is apparent in this list, let us change the left column of indicated divisions by using exponents. The result would then be as follows:

$$10^2/10^1 = 10^1$$
$$10^3/10^1 = 10^2$$
$$10^4/10^1 = 10^3$$
$$10^5/10^2 = 10^3$$
$$10^6/10^2 = 10^4$$

In each division, the difference of the exponents on the left is the same as the exponent on the right. Again, this result is not accidental. It also follows from the definition of a positive integral exponent. For example, in the fourth division above, 5 factors of 10 are divided by 2 factors of 10; therefore, there are only 3 factors of ten left. This law of division also follows from an algebraic rule. Symbolically, the rule is

$$x^n/x^m = x^{n-m}$$

In words, the rule says that if two numbers *with the same base* are divided, the quotient is a number having the same base, but with an exponent which is the difference of the exponents of the dividend and the divisor. By this rule,

$$8^6/8^2 = 8^{6-2} = 8^4$$

But $8^9/5^4$ could not be evaluated by this rule, since the bases would not be the same.

Using the above rule, what would 10^9 divided by 10^4 equal? The base of 10 is the same for dividend and divisor, therefore the exponents can be subtracted. Performing the division gives

$$10^9/10^4 = 10^{9-4} = 10^5$$

PROBLEMS

1. Carry out the following divisions.
   a. $10^8/10^3$ b. $10^5/10^4$ c. $10^{23}/10^5$ d. $10^{14}/10^{10}$

2. Change each of the following numbers to power of ten form, and perform the divisions.
   a. 1000/10   b. 10,000,000/1000   c. 10,000,000,000/10,000
   d. 1,000,000,000/1,000,000

3-4 The Negative Exponent

We now have two rules for using exponents on the base of 10. The rules can be stated symbolically by

$$(10^n)(10^m) = 10^{n+m}$$

for multiplication, and by

$$10^n/10^m = 10^{n-m}$$

for division. However, something interesting happens with certain kinds of quotients. Look at the following list of indicated divisions.

$$\frac{10}{100} = \frac{10}{(10)(10)} = \frac{1}{10} = 1/10^1$$

$$\frac{10}{1000} = \frac{10}{(100)(10)} = \frac{1}{100} = 1/10^2$$

$$\frac{10}{10,000} = \frac{10}{(1000)(10)} = \frac{1}{1000} = 1/10^3$$

$$\frac{1000}{100,000} = \frac{1000}{(100)(1000)} = \frac{1}{100} = 1/10^2$$

$$\frac{100}{1,000,000} = \frac{100}{(10,000)(100)} = \frac{1}{10,000} = 1/10^4$$

Now examine the same list when the quotients in the left column are expressed with exponents.

$$10^1/10^2 = 1/10^1$$
$$10^1/10^3 = 1/10^2$$
$$10^1/10^4 = 1/10^3$$
$$10^3/10^5 = 1/10^2$$
$$10^2/10^6 = 1/10^4$$

Look for a pattern involving exponents. The exponent on the right is the difference of the exponents of the quotient on the left, but in a different way than it was for the divisions performed when the exponents of the numerator were larger than the exponents of the denominator.

Consider what would happen if the rule for division developed earlier were used in the above list of quotients. The rule is

$$10^n/10^m = 10^{n-m}$$

Thus, the quotients become

$$10^1/10^2 = 10^{1-2} = 10^{-1}$$
$$10^1/10^3 = 10^{1-3} = 10^{-2}$$
$$10^1/10^4 = 10^{1-4} = 10^{-3}$$
$$10^3/10^5 = 10^{3-5} = 10^{-2}$$
$$10^2/10^6 = 10^{2-6} = 10^{-4}$$

If the rule of division is to be applied in those cases where negative exponents are the result, then by comparing this list of divisions with the previous list, we can determine what meaning *we* must give to the negative exponent. Examine that comparison. We must have

$$1/10^1 = 10^{-1}$$
$$1/10^2 = 10^{-2}$$
$$1/10^3 = 10^{-3}$$
$$1/10^2 = 10^{-2}$$
$$1/10^4 = 10^{-4}$$

The pattern is clear; the exponents are the same, except for sign. The general relationship for powers of 10 must be that

$$1/10^n = 10^{-n}$$

Notice that we have not proved that this relationship is true, nor could we do so. We are making a definition of the negative exponent in such a way as to permit us to use our division rule in every case. We have decided what the negative exponent is going to mean. The general relationship for any base would be, by definition,

$$1/x^n = x^{-n}$$

It may not yet be clear how negative exponents can be helpful in computation. But examine the following list of numbers:

$$0.1 \quad = 1/10 \quad = 1/10^1 = 10^{-1}$$
$$0.01 \quad = 1/100 \quad = 1/10^2 = 10^{-2}$$
$$0.001 \quad = 1/1000 \quad = 1/10^3 = 10^{-3}$$
$$0.0001 \quad = 1/10{,}000 \quad = 1/10^4 = 10^{-4}$$
$$0.00001 = 1/100{,}000 = 1/10^5 = 10^{-5}$$

The first and last columns show how small numbers can be expressed using negative exponents. A simple method of determining the exponent is to count how many places the decimal point would have to be moved in order to get the number 1. For example, in the number 0.00000001, the decimal point would have to be moved 8 places to the right to get the number 1. Thus $0.00000001 = 10^{-8}$.

PROBLEMS

1. Express each of the following numbers in power of ten form.
   a. 0.1   b. 0.00000001   c. 0.001   d. 0.00001

2. Using power of ten notation, express each of the following quantities in terms of the unit indicated.
   a. 1 μ in m   b. 1 nsec in sec   c. 1 mm in m   d. 1 cm in m

3. Rewrite each of the following numbers without using exponents.
   a. 10^{-7} b. 10^{-13} c. 10^{-2} d. 10^{-23}

3-5 The Zero Exponent

With what has been developed so far in this chapter, it is now possible to construct a list of large and small numbers as follows:

$$100{,}000 = 10^5$$
$$10{,}000 = 10^4$$
$$1000 = 10^3$$
$$100 = 10^2$$
$$10 = 10^1$$
$$1 = ?$$
$$0.1 = 10^{-1}$$
$$0.01 = 10^{-2}$$
$$0.001 = 10^{-3}$$
$$0.0001 = 10^{-4}$$
$$0.00001 = 10^{-5}$$

By studying this list of numbers, with the corresponding power of ten form, the way of expressing these kinds of large and small numbers becomes apparent.

There is one number in the list above that is uncertain. What exponent should be used on the base ten in order to give the number 1? From the order of the exponents in the list, one might guess it to be 10^0, but perhaps a better reason is needed before selecting such a strange exponent.

One fact about arithmetic is obvious. Any number divided by itself is 1. Thus we divide numbers by themselves, as in the following list:

$$1 = 100/100 = 10^2/10^2 = 10^{2-2} = 10^0$$
$$1 = 1000/1000 = 10^3/10^3 = 10^{3-3} = 10^0$$
$$1 = 10{,}000/10{,}000 = 10^4/10^4 = 10^{4-4} = 10^0$$

Previously we used the rule for division,

$$10^n/10^m = 10^{n-m}$$

when n was larger than m, and when n was smaller than m. In the latter case, we had to invent negative exponents to use the rule. In the present case, as the list above indicates, if this division rule is to be valid for all integral exponents including zero, we must have $10^0 = 1$ as a definition of the zero exponent. Again, we have not proved that $10^0 = 1$; we have made this a definition to permit the use of the division rule in this type quotient. The definition can then be made general by stating it as

$$x^n/x^n = x^{n-n} = x^0 = 1$$

This equation, $x^0 = 1$, is the *definition* of the zero exponent.

PROBLEMS

1. Find the value of each of the following quantities.
    a. 12^0 b. 4^0 c. $5^0/2$ d. $8^0 - 4^0$

2. Carry out each of the following multiplications, showing the final answer as a number without exponents.
    a. $(10^8)(10^{-8})$ b. $(10^{12})(10^{-12})$ c. $(10^0)(10^0)$

3-6 Expressing Numbers in Powers of Ten

By means of the rules of multiplication and division,

$$(x^n)(x^m) = x^{n+m} \qquad x^n/x^m = x^{n-m}$$

and the definitions of negative exponents and the zero exponent,

$$1/x^n = x^{-n} \qquad x^0 = 1$$

using the base of 10 we can write large and small numbers with exponents and are able to multiply and divide such numbers easily.

[§ 3-6] *Expressing Numbers in Powers of Ten* **35**

How to express numbers like 1,000,000 and 0.000001 using powers of ten has already been shown, but how can powers of ten be used to express such numbers as 1,423,000,000 or 0.00000323? These are, after all, the kinds of numbers with which we must make computations.

In order to use powers of ten for large or small numbers, one must decide how he wants the numbers expressed. As a matter of convention in this textbook, a number will be expressed between 1 and 10 times a power of ten. As an example, what would be the correct expression for 14? This number is

$$(1.4)(10) = 1.4 \times 10^1$$

The $\times$ is used here to mean multiplication because it is not likely to be confused as a variable, and this method is used rather extensively by people in science and technology.

The basic problem in expressing a number in power of ten form is to factor it into two numbers, the first a number between 1 and 10, and the second some number which has an integral power of ten. The number 278,000 can be factored into 27,800 $\times$ 10, 2780 $\times$ 100, 278 $\times$ 1000, 2.78 $\times$ 100,000, 0.278 $\times$ 1,000,000, etc. One could keep going with this process indefinitely, but we are interested only in the factors with the first between 1 and 10. Thus, we would use the factors 2.78 $\times$ 100,000. But 100,000 = 10^5, therefore we can write 278,000 as 2.78 $\times$ 10^5.

Consider the following list of numbers which have been put into the correct power of ten form:

$$12. = 1.2 \times 10^1$$
$$480. = 4.8 \times 10^2$$
$$7,320. = 7.32 \times 10^3$$
$$29,000. = 2.9 \times 10^4$$
$$335,000. = 3.35 \times 10^5$$
$$5,698,200. = 5.6982 \times 10^6$$

A careful inspection reveals that the exponent on the power of ten is the same as the number of places to the left that the decimal point must be moved to get a number between 1 and 10 from the original number. If one understands the basic principle of exponents and how factorization leads to the correct form for power of ten expression, memorization and use of this present rule is acceptable. But merely to memorize the rule without an understanding of it is dangerous, since, if it is forgotten, one would be unable to derive it again as needed. The use of such rules is encouraged, but only if their origin is understood.

The above discussion shows how large numbers can be expressed in power of ten form. But what about small numbers? How would a number like 0.0000054 be expressed in this form? It must be factored into two numbers, one between 1 and 10, and the other some integral power of ten. Factoring such numbers can be done by forming a simple decimal multiplied by the number, expressed between 1 and 10. The number 0.0000054 can be factored to (0.000001)(5.4) just by counting zeros. But, as already shown, $0.000001 = 10^{-6}$, thus the number becomes 5.4×10^{-6} in power of ten form.

Examine the following list of numbers which have been put into the correct power of ten form.

$$0.12 = 1.2 \times 10^{-1}$$
$$0.048 = 4.8 \times 10^{-2}$$
$$0.00732 = 7.32 \times 10^{-3}$$
$$0.00029 = 2.9 \times 10^{-4}$$
$$0.0000335 = 3.35 \times 10^{-5}$$
$$0.000005698 = 5.698 \times 10^{-6}$$

The absolute value of the exponent in each power of ten in the right hand column is the number of places the decimal point must be moved to the right to get a number between 1 and 10. Again, if the principle of factorization is understood, the use of the above rule is acceptable. How would the rule be applied to express the number 0.00000382 in power of ten form? To get a number between 1 and 10, the decimal point must be moved 6 places to the right. Thus the desired form would be 3.82×10^{-6}.

PROBLEMS

1. The distance from the earth to the sun is 314,000,000,000 m. How far is this, expressed in power of ten form?

2. The mass of the earth is 5,960,000,000,000,000,000,000,000 kg. Express this mass in power of ten form.

3. The charge on the electron is 1.60×10^{-19} coulombs. Express this number without using exponents.

4. The mass of the proton is 0.000000000000000000000000000167 kg. Express this number in power of ten form.

5. Express each of the following numbers in power of ten form.
   a. 0.00000136   b. 10057000   c. 0.00200056   d. 1

3-7 Multiplication of Large and Small Numbers

The ability to express numbers in power of ten form greatly reduces the difficulty of doing arithmetic using large and small numbers. The purpose of learning to express numbers in this form is to make computation easier. Before considering such problems, however, one must review an elementary algebraic principle involving the multiplication of numbers.

This algebraic principle states simply that any factor in a group of factors is used only once. For example, in the problem

$$(2)(3)(6)(5) = ?$$

the factor 2 is used only once; it does not multiply every factor present. Of course the order of multiplication does not change the result. Thus

$$(2)(3)(6)(5) = (6)(6)(5) = (36)(5) = 180$$

Or multiplication could be performed in a different order, giving

$$(2)(3)(6)(5) = (2)(3)(30) = (2)(90) = 180$$

with the same result as before.

In the use of power of ten notation, the above principle permits the separate multiplication of powers of ten, and of the numbers preceding the powers of ten. For example, 2×10^6 multiplied by 3×10^4 would be

$$(2 \times 10^6)(3 \times 10^4) = (2)(3)(10^6)(10^4) = 6 \times 10^{10}$$

In problems involving more than two factors, a similar technique is used. All factors between 1 and 10 are grouped together, and all powers of ten are grouped together, the arithmetic being performed on each group separately. For example, consider the product of 3×10^7, 5×10^{22}, 4×10^3 and 7×10^{13}. First, one would group the factors 3,5,4 and 7, then the powers of ten. They would appear as follows:

$$(3)(5)(4)(7)(10^7)(10^{20})(10^3)(10^{13})$$

The multiplication of the factors $(3)(5)(4)(7)$ gives

$$(15)(4)(7) = (60)(7) = 420$$

Next, the exponents of the powers of ten must be added to find the exponent to be placed on 10 in the result. The sum gives

$$7 + 22 + 3 + 13 = 45$$

Thus we have an answer of

$$420 \times 10^{45}$$

But notice that this number is not in the desired form of a number between 1 and 10 times a power of ten. It is a simple matter to change 420 to 4.2×10^2. Thus we have two more factors of ten in that part of our answer. The final result is 4.2×10^{47}.

In the above examples, only large numbers were multiplied together. But very often, both large and small numbers must be multiplied. Fortunately, the technique is identical if the rule for multiplying numbers having the same base is followed carefully. Consider the following problem:

$$(2 \times 10^{-9})(3 \times 10^{12}) = ?$$

Grouping factors, we have

$$(2)(3)(10^{-9})(10^{12}) = 6 \times 10^{-9+12} = 6 \times 10^3$$

When the exponents are added, the minus 9 and the plus 12 add to give a plus 3. In the case of many factors of large numbers (positive exponents) and of small numbers (negative exponents), the algebraic sum of the exponents is what must be placed on the base ten of the answer. As an example, look at the following problem:

$$(2 \times 10^{-23})(3 \times 10^{18})(5 \times 10^6)(9 \times 10^{-6}) = ?$$

Properly grouping factors gives

$$(2)(3)(5)(9)(10^{-23})(10^{18})(10^6)(10^{-6})$$

But this is the same as

$$270 \times 10^{-23+18+6-6} = 270 \times 10^{-29+24} = 270 \times 10^{-5}$$

Again, this answer is not quite in the desired form, since 270 can be factored to 2.7×10^2. Using this result, the above answer becomes

$$(2.7 \times 10^2)(10^{-5}) = 2.7 \times 10^{2-5} = 2.7 \times 10^{-3}$$

As a general procedure, to multiply factors expressed in power of ten form, group the factors which are between 1 and 10 together and the powers of ten together; then carry out the multiplication of the first group, re-expressing that result in power of ten form. By an algebraic sum of the exponents of the powers of 10, an exponent can be found to be placed on that base. Finally, an adjustment to the latter power of ten result is made to get the answer in the form of a number between 1 and 10 times an integral power of ten. These words simply describe the process which has been used in the preceding examples.

PROBLEMS

1. Carry out the following multiplications using power of ten notation.
   a. $(5 \times 10^3)(6 \times 10^{87})$ b. $(3 \times 10^{-9})(2 \times 10^{-7})$
   c. $(8 \times 10^{-4})(7 \times 10^{12})$ d. $(6 \times 10^{-9})(3 \times 10^{-4})$

2. Express each of the following factors in power of ten form; then carry out the indicated multiplications.
   a. $(0.000004)(0.0009)$ b. $(8,000,000,000)(0.00007)$
   c. $(0.000000000000000005)(0.0000000004)$
   d. $(3,000,000,000)(90,000)$ e. $(0.0000000008)(2,000,000,000,000)$

3-8 Division of Large and Small Numbers

The process of division of large and small numbers involves procedures very similar to that of multiplication. Factors in numerator and denominator must first be grouped as in the process of multiplication. Then numbers are divided as indicated, with those having powers of ten divided by the rule for division, $10^n/10^m = 10^{n-m}$.

As a simple example, consider the division of 6×10^8 by 3×10^5. This division would appear as

$$(6 \times 10^8)/(3 \times 10^5) = (6/3) \times (10^8/10^5) = 2 \times 10^{8-5} = 2 \times 10^3$$

A more complicated problem might contain several factors in the numerator and denominator:

$$\frac{(3 \times 10^{12})(5 \times 10^7)(4 \times 10^8)}{(2 \times 10^5)(3 \times 10^4)}$$

In this problem, multiplications in the numerator and denominator can be separately computed, as shown previously. This process gives a result of

$$(60 \times 10^{27})/(6 \times 10^9)$$

a simple division which gives

$$(60/6) \times (10^{27}/10^9) = 10 \times 10^{18} = 10^{19}$$

for the quotient.

The type of problem which causes much difficulty involves division with negative exponents present. How would 10^7 be divided by 10^{-3}?

One would have $10^7/10^{-3}$, and, by the rule for division mentioned above, the result would be

$$10^{7-(-3)} = 10^{7+3} = 10^{10}$$

This problem is usually done incorrectly, resulting instead in an answer like 10^4. But a little thought about the numbers helps one see why the larger number, 10^{10}, is more reasonable. In dividing 10^7 by 10^{-3}, we are dividing a large number (10,000,000) by a small number (0.001). Thus we should expect that this small number divided into a large number should be a very large number, as is 10^{10} (10,000,000,000). The difficulty arises because some students never really understood algebraic subtraction. To subtract, change the sign of the subtrahend and perform the process of algebraic addition. For example, in subtracting (-9) from $(+6)$, we would write $(+6) - (-9)$, where (-9) is the subtrahend. Changing the sign of the subtrahend and adding,

$$(+6) - (-9) = (+6) + (+9) = 6 + 9 = 15$$

With this reminder from elementary algebra, divisions involving negative exponents should cause a minimum of difficulty.

Consider this more complicated problem:

$$\frac{(5 \times 10^{-6})(4 \times 10^{13})(7 \times 10^7)}{(2 \times 10^{-24})(10^7)(10^{-5})}$$

When numerator and denominator multiplications are carried out first, one has

$$(140 \times 10^{14})/(2 \times 10^{-22})$$

which becomes

$$(140/2) \times (10^{14}/10^{-22}) = 70 \times 10^{14-(-22)} = 70 \times 10^{14+22}$$
$$= 70 \times 10^{36} = 7.0 \times 10^{37}$$

By studying this example carefully, one can better understand the general process of division of large and small numbers.

PROBLEMS

1. Carry out the following divisions.
   a. $6 \times 10^7/3 \times 10^5$ b. $8 \times 10^{-8}/4 \times 10^{-5}$ c. $9 \times 10^3/2 \times 10^{-22}$
   d. $4 \times 10^{-23}/8 \times 10^{17}$

[§ 3-9]   Roots and Powers   41

2. Change all numbers below to power of ten form and perform the indicated divisions.
   a. 0.000008/0.000004   b. 0.0000009/3,000,000
   c. 8,000,000,000/0.0000004
   d. 60,000,000,000,000/300,000,000,000,000

3. Change the following numbers to power of ten form and carry out the indicated operations.
$$\frac{(0.00004)(8,000,000,000,000)(0.000000005)}{(2000)(0.004)(0.0000008)(400,000,000,000)}$$

4. The charge on the electron is 1.60×10^{-19} coulombs, and the mass of the electron is 9.1×10^{-31} kg. What is the ratio of charge to mass (divide charge by mass) for the electron?

3-9 Roots and Powers

Power of ten expression is especially useful in finding roots of numbers, or in raising numbers to powers. But the method of expression depends upon what operation is to be performed on the number.

Raising numbers to powers is relatively simple. The algebraic rule for such an operation involving a number times a power of ten raised to some power is:

$$(N \times 10^n)^m = N^m \times 10^{(n)(m)}$$

This rule states that both the number N and the number 10^n are raised to the power m. To raise 10^n to the m power, m is multiplied times n to give the exponent to be placed on the base 10. As an example, the quantity

$$(2 \times 10^5)^8 = 2^8 \times 10^{(5)(8)} = 2^8 \times 10^{40} = 256 \times 10^{40} = 2.56 \times 10^{42}$$

The value of 2^8 must be found by multiplying the 8 factors of 2 to get 256.

Finding roots is somewhat more difficult than raising numbers to powers. Consider first the process of extracting square roots. A helpful fact to recall is the algebraic definition of the exponent 1/2. This exponent is defined by the equation

$$\sqrt{x} = x^{1/2}$$

This definition of the fractional exponent 1/2 makes it possible to treat it like any other exponent, using the rules of algebra. For example, using the rule $(x^n)^m = x^{(n)(m)}$,

$$\sqrt{25} = (25)^{1/2} = (5^2)^{1/2} = 5^{(2)(1/2)} = 5^1 = 5$$

Because the square root is the same as the power 1/2, any number in power of ten form for which a square root is to be found must have its power of ten exponent as an even integer. The reason the power of ten must be even is that 1/2 of an even integer is also an integer, and those are the kinds of exponents we need. For example, $\sqrt{3.6 \times 10^5}$ cannot be evaluated in this form because the exponent 5 is not an even integer. But the radicand, 3.6×10^5, can be changed to 36×10^4, which does have an even exponent on the 10. Now we have $\sqrt{36 \times 10^4}$, which, by the equation defining the exponent 1/2, becomes

$$(36 \times 10^4)^{1/2} = (6^2 \times 10^4)^{1/2} = 6^{(2)(1/2)} \times 10^{(4)(1/2)}$$
$$= 6^1 \times 10^2 = 6 \times 10^2 = 600$$

To find cube roots, fourth roots, etc., definitions are made using fractional exponents, so that the rules of algebra can be used with these roots also. For example, the exponent 1/3 is defined by the equation

$$\sqrt[3]{x} = x^{1/3}$$

This definition indicates that if we are to find a cube root of a number expressed in power of ten form, the exponent on the 10 must be divisible by three. Similarly, if we were finding a fourth root, the exponent on the 10 would have to be divisible by four.

PROBLEMS

1. Express each of the following as numbers between 1 and 100 times an even power of ten.
   a. 0.00000000049   b. 0.0000000049   c. 64,000   d. 8,100,000,000
   e. 0.000000000000000036

2. By writing the following numbers between 1 and 100 times an even power of ten, find their square roots.
   a. 3,600,000,000,000,000   b. 0.0000000009
   c. 0.000000000000000000000004
   d. 6,400,000,000,000,000,000,000,000

3. Find the square roots of the following numbers.
   a. 4.9×10^5 b. 9×10^{-28} c. 6.4×10^{-13} d. 4×10^{38}

3-10 General Computations

With the ability to express numbers in power of ten form and to carry out operations of multiplication and division, the process of com-

[§ 3-10]   *General Computations*   **43**

putation becomes simple. The kinds of numbers used in examples so far in this chapter have given integral results. No difficult arithmetic has been included. The numbers have been the kind for which computation could be done instantly.

The typical problem that occurs in science and technology involves large and small numbers, with sums and differences, and multiplications and divisions. Sometimes roots and powers are included. Such problems can appear tediously long, yet by the use of power of ten notation, they can be reduced to simpler problems which could be easily solved if the arithmetic could be done. In the following chapter, it will be shown how the slide rule can do the arithmetic when problems are stated in power of ten form.

4 Slide Rule Computation

The ability to use power of ten notation enables one to solve complex arithmetic problems as easily as his skill in doing simple computation permits. Of course, one should not waste time computing with pencil and paper if the task can be accomplished in some other way much more quickly. The slide rule can well serve this purpose.

4-1 The Slide Rule

The slide rule has a basically simple construction. As shown in Figure 4-1, it consists of the upper and lower fixed parts called the *body*. The *slide* is between the above two parts, and moves in grooves cut in the body. The *indicator* (cursor) is made of a sliding window with a hairline.

[§ 4-2] *Slide Rule Scales* 45

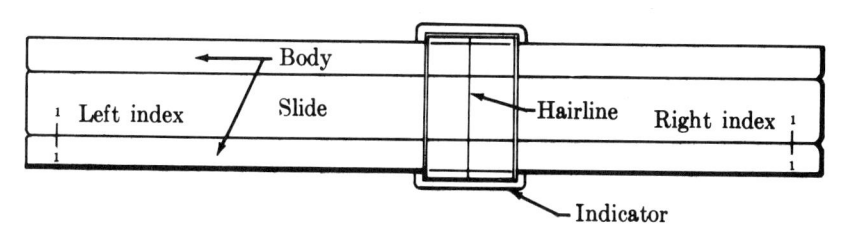

Figure 4-1. *The parts of a slide rule.*

Scales are usually cut into the body and slide, and marked off in black. For the operations of multiplication and division, scale C on the lower edge of the slide, and the adjacent scale D on the body can be used. For square roots, scales D and A on the body can be used.

One should examine his own slide rule carefully, learning to associate the words *slide, hairline, indicator, body, right index, left index,* C *scale,* D *scale,* and A *scale* with the correct part or location on the instrument.

4-2 Slide Rule Scales

An examination of the C scale on the slide and the D scale on the body of a slide rule shows that these two scales are identical when placed with indexes properly aligned. It is seen immediately that the numbers marked on these scales are not evenly spaced. Look at Figure 4-2. This non-linearity is what permits the slide rule to be used for multiplication and division. The spacing of the scale marks and numbers is done in such a way as to convert each number to an exponent. When numbers are multiplied on a slide rule, two lengths of the slide rule are added together; thus the user is adding exponents. When numbers are divided, one length is subtracted from another length, expo-

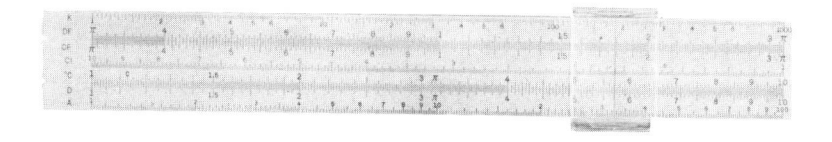

Figure 4-2. *Slide rule scales.*

nents thereby being subtracted. It is therefore possible to multiply and divide numbers on a slide rule by adding or subtracting distances along the scales. By reading the scales, one automatically converts from exponents back to numbers.

The other slide rule scale which will be considered is the A scale. Figure 4-3 shows the A scale next to the D scale. Notice that the A scale consists of two D scales which have been compressed into the same length as the D scale. With the A and D scales it is possible to read squares or square roots directly. Again, the principle involves exponents. Because the square root is the same as the exponent 1/2, the A scale cycle is made only half as long, thus requiring two full D scales along the length of the rule. Most slide rules have another scale, the K scale, which has three D-type scales compressed along the length of the rule. When used with the D scale, this K scale would give cubes or cube roots. In the same way, four D scales compressed along the rule would provide for fourth powers or fourth roots.

Figure 4-3. *The A and D scales.*

On the more elaborate slide rules, a number of other scales are present. But the greatest use of the slide rule is in multiplication, division, and taking roots; thus the other scales will not be considered here. Once you gain speed and confidence in the elementary operations with your own slide rule, you can readily learn to use other scales as needed.

4-3 Reading the Scales

The ability to use power of ten notation in computations easily enables one to locate the decimal point in an arithmetic result. Through this technique, the slide rule can be used to do arithmetic without concern for the location of a decimal point. The slide rule operates on digits only. It does not locate decimal points.

The C or D scale of a slide rule is divided into uneven divisions from one to ten. Figure 4-4 shows how the spaces between consecutive numbers become smaller as the numbers increase from 1 to 10. This non-linearity of scale divisions exists because the scale is exponential.

[§ 4-3]   Reading the Scales   47

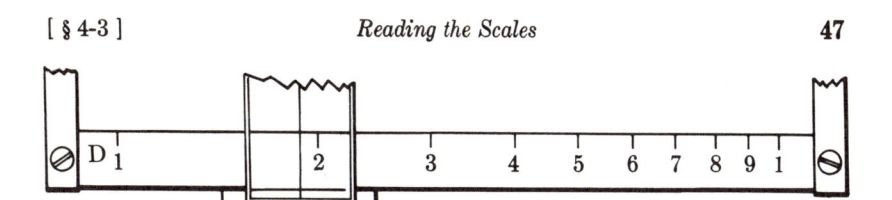

Figure 4-4.   *Scale divisions are uneven.*

As shown in Figure 4-5, the C and D scales are subdivided, but not in the same way all along the slide rule. Between 1 and 2 are the numbers 1 through 9. These numbers correspond with 11 through 19 respectively. The number 6 between 1 and 2 represents 16, the number 3 between 1 and 2 represents 13, and so forth. Each of these ten subdivisions between 1 and 2 is further subdivided into 10 smaller divisions, with the fifth mark in each case being slightly longer than the other ones. Because of this type of subdivision between 1 and 2, it is possible to read three digits directly, and make a good estimate of a fourth digit. For example, in Figure 4-6 the hairline of the indicator

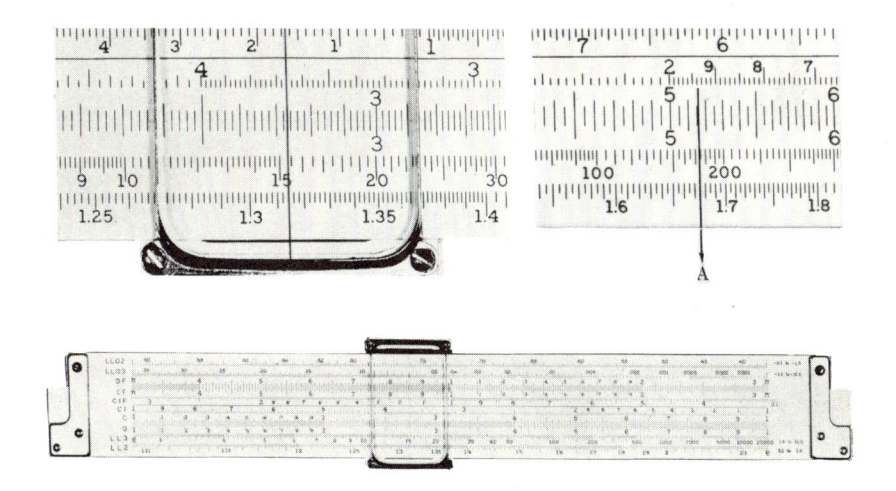

Figure 4-5.   *Read the D scale.*

48 *Quantitative Aspects of Science and Technology* [§ 4-3]

crosses the D scale between main divisions of 1 and 2, and between subdivisions of 2 and 3. In that subdivision, the hairline crosses between the fifth and sixth mark at about 52 or 53. Thus, the hairline reads 1252 or 1253. Let us say that it reads 1253. Notice that we are not saying that this reading is 1.253, 12.53, or 125.3, or even 1253 as a number. We are merely saying that the reading gives the digits 1253, and that is our only concern at this time. Later, we can be concerned with locating a decimal point in a given problem.

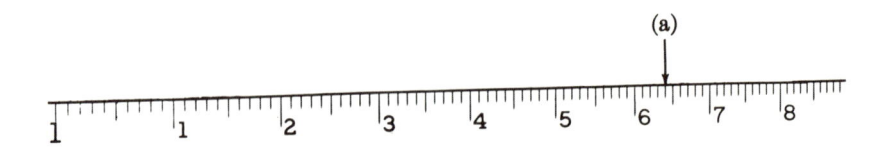

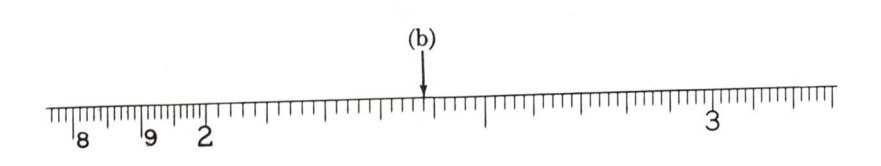

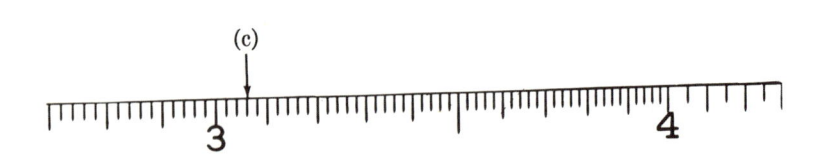

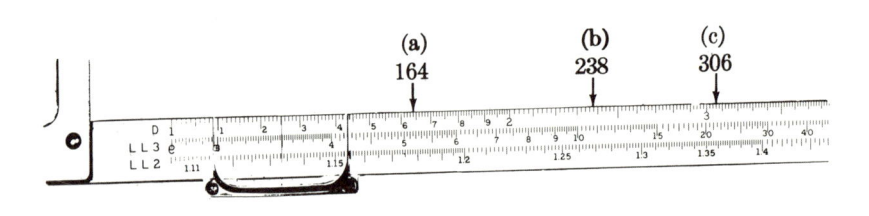

Figure 4-6. *The D scale readings are (a) 164, (b) 238, and (c) 306.*

[§ 4-3]   *Reading the Scales*   **49**

The subdivisions on scales C and D between the main divisions of 2 and 3 and 3 and 4 are similar in each case, but quite different from those between 1 and 2. Examine those subdivisions on the slide rule of Figure 4-5. None of the subdivisions in these regions are marked with digits, although there are 10 subdivisions in each case, which are then further subdivided. But the smaller subdivisions are in groups of five rather than ten. This means that the smallest subdivisions between 2 and 3 and 3 and 4 each represent units of two. To understand what this means, look at the hairline of the indicator in Figure 4-5. It crosses between the main scale numbers of 2 and 3. By counting the longer subdivision marks, one can see that the hairline crosses between the seventh and eighth mark. It is past the first smallest mark within that subdivision, but not quite to the second mark. This does not mean that the last digit is between 1 and 2, but instead, because there are only 5 spaces, it means that the last digit is between 2 and 4. Because the hairline appears to cross near the middle of that second space, we can say that the last digit is 3. Thus the reading would be 273.

The main scale divisions from 4 to 5, 5 to 6, 6 to 7, 7 to 8, 8 to 9, and 9 to the right index of 1 are each subdivided in the same way. As shown in Figure 4-5, these spaces are subdivided into 10 spaces, each of which has a center mark. A line in Figure 4-5 shown with the letter *A* crosses the C and D scales between 5 and 6. The line crosses the subdivisions halfway between the first and second. The reading is therefore 515.

Once you have examined the C and D scales carefully, counting the various subdivisions and practicing making readings, you will soon be able to look at an index or indicator hairline and quickly take a correct scale reading. Of course this ability is essential before one can begin multiplying or dividing on the slide rule.

The D scale of Figure 4-6 shows three arrows, each pointing to scale marks. Reading (a) is between the main scale marks of 1 and 2 and between sub-marks of 6 and 7. By counting the smallest sub-marks, the reading is 164. Reading (b) is between main scale marks of 2 and 3 and between the third and fourth sub-marks. Because this smallest subdivision is in five parts, and the arrow points to the fourth mark, the scale reading is 8/10 of the way across. Thus the scale reading is 238. Reading (c) is between main scale marks of 3 and 4, but not quite to the first subdivision mark. Again, the smallest subdivision is in five parts, the arrow pointing to the third mark. The reading would therefore be 306. If you can make each of these kinds of readings without difficulty, you are ready to learn the operations of multiplication and division with the slide rule.

PROBLEMS

1. Read the digits of each of the following examples below the C scale index.

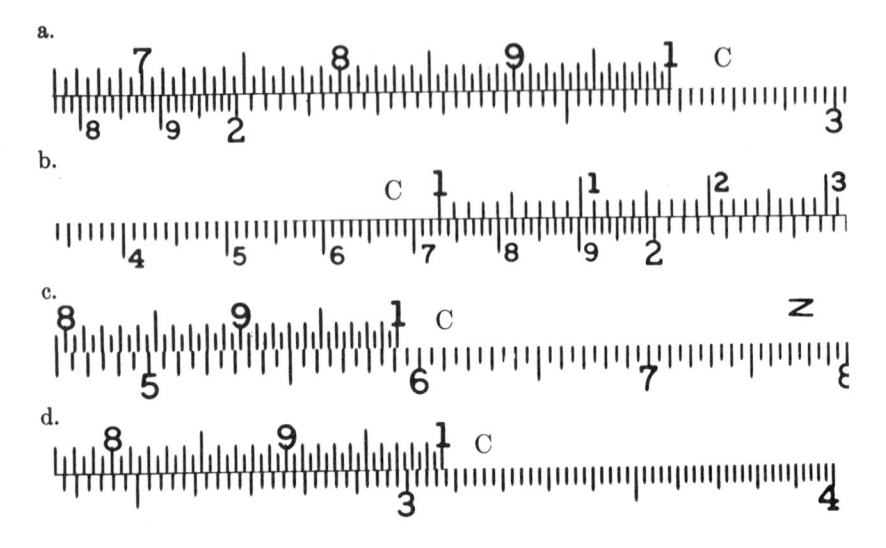

   a.

   b.

   c.

   d.

2. Read the digits of each of the following examples where the hairline crosses the D scale.

   a. D

   b. D

   c. D

   d. D

4-4 Multiplication

The process of multiplication with the slide rule is based on the principle of adding exponents. With powers of ten, this principle appeared as $(10^n)(10^m) = 10^{n+m}$. For the C and D scales of the slide rule, each scale is exponentially divided in an identical way. This means, essentially, that each scale can be considered as some base on which the scale marks represent numerical values, and the distances along the scales represent exponents. Because distances represent exponents, to multiply numbers on the slide rule two distances must be added, one on the C scale and the other on the D scale. Then the product can be read directly as the scale number corresponding to the sum of those distances.

To illustrate this principle, the numbers 2 and 3 are multiplied in Figure 4-7. To multiply these two numbers, a scale distance from 1 to 2 and a scale distance from 1 to 3 must be added. Then, the scale number that corresponds with the sum of those two distances is the correct product of the numbers. The way to add these distances on the slide rule is to place the left index of C directly above the number 2 on the D scale. Then the indicator is moved to where the hairline crosses the number 3 on the C scale. From the left index of the D scale to where the hairline crosses the D scale is the sum of the two scale distances; the point where the hairline crosses the D scale is therefore the correct product of the numbers 2 and 3. As shown in Figure 4-7, that product is, as expected, equal to 6.

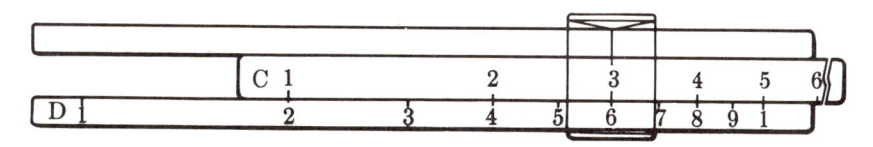

Figure 4-7.   *Two times three is six.*

As mentioned previously, the slide rule C and D scales do not contain numbers with decimal places. They contain only digits, and the user must locate decimals himself. However, one might consider the scales as being marked off from 1 to 10, as long as one uses the left index of the C scale in multiplication. Eventually, however, you will have to multiply two numbers for which the product does not fall on the D scale. For example, if the numbers 3 and 5 are multiplied, using the left index of the C scale as before, a problem arises. If the left index is placed on 3 of the D scale, the indicator cannot be moved to the 5 of the C scale, since that number is then out beyond the right end of the slide rule. It appears that another D scale attached at the right end of the rule is needed.

It would not be very convenient to attach another D scale to the slide rule, but there is a way of using the slide rule as though it had another scale attached. The right index can be used. In effect, the use of the right index provides an imaginary extra D scale at the left end of the slide rule. Figure 4-8 illustrates the process of multiplication of the numbers 3 and 5 using the right index. The right index of the C scale is placed over the 5 of the D scale. Then the indicator is placed so that the hairline crosses the 3 on the C scale. The product of 3 and 5 would then be where the hairline crosses the D scale, in this case at the digits 15.

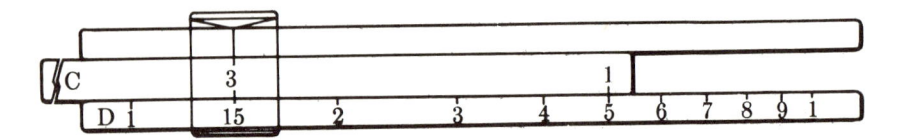

Figure 4-8. *Three times five is fifteen.*

From the examples already discussed, one can describe the steps needed to perform a multiplication with the slide rule. First, it must be determined which C scale index is to be used by roughly figuring whether the product will be greater than 10 or less than 10 when the factors are considered as numbers between 1 and 10. For example, in multiplying the digits 3482 and 843, since 3.482 times 8.43 will be somewhere around 24 or 25, the right C scale index must be used. When one has determined which C scale index is to be used, he places that index over the D scale reading of either of the two factors involved. Then the indicator is moved to where the hairline crosses the other factor on the C scale. Where the hairline then crosses the D scale is the correct product.

Up to this point, we have discussed multiplication of digits without reference to decimal points. In actual problems, it is important to know where the decimal point is to be located. By the use of powers of ten and a little rough computation, it is possible always to keep track of where the decimal point belongs.

When we have learned the basic process of multiplication of digits on the slide rule, it is not difficult to use the slide rule in multiplication of numbers with decimal points. To keep account of those decimal points, we must use powers of ten and make a few approximate mental computations.

[§ 4-4]   *Multiplication*   53

The first step in any multiplication of numbers is to express each factor as a number between 1 and 10 times some power of ten. Then the slide rule is used to multiply the numbers between 1 and 10, the powers of ten being multiplied by adding exponents.

Consider the following example. It is desired to multiply the numbers 1,435,000 and 0.00151. First they must be expressed as numbers between 1 and 10 times some power of ten. The first factor is the same as 1.435×10^6, and the second factor is the same as 1.51×10^{-3}. Showing the multiplication, we have

$$(1.435 \times 10^6)(1.51 \times 10^{-3}) = (1.435)(1.51) \times 10^3$$

The exponent on the power of ten is determined by adding the two exponents 6 and −3 to get 3. The multiplication of the numbers 1.435 and 1.51 is done on the slide rule of Figure 4-9. Because the product is, from a mental computation, somewhere around 2 or 3 (a result certainly less than 10), we know to use the left C scale index. Thus the left C scale index is placed over the digits 1435 on the D scale. The indicator is moved to where the hairline crosses the C scale at 151. Then the product is where the hairline crosses the D scale. In this case, the product is between 216 and 218, say 217. By our mental arithmetic we know that the product of 1.435 and 1.51 must be a number somewhere around 2 or 3, and our slide rule shows the correct digits as 217. Thus the correct product must be 2.17. The final answer to our original problem must then be 2.17×10^3, a number which could be left in this form, or if written without exponents, would appear as 2170.

PROBLEMS

1. With the slide rule, multiply the following sets of two factors. Express answers in powers of ten form.
    a. (367,000)(275)   b. (4,500,000)(75,000)
    c. (78,000,000)(0.0000000000000443)   d. (128,000)(0.0000000051)
    e. (0.000398)(0.000000448)   f. (0.0098)(0.000088)

2. With one series of slide rule operations, multiply the following sets of three factors. Express answers in power of ten form.
    a. (0.00049)(5,000,000)(0.000081)
    b. (1,070,000)(475,000)(706,000,000)
    c. (0.0000077)(0.000429)(0.0000104)
    d. (50)(0.0000000000000924)(385,000,000)

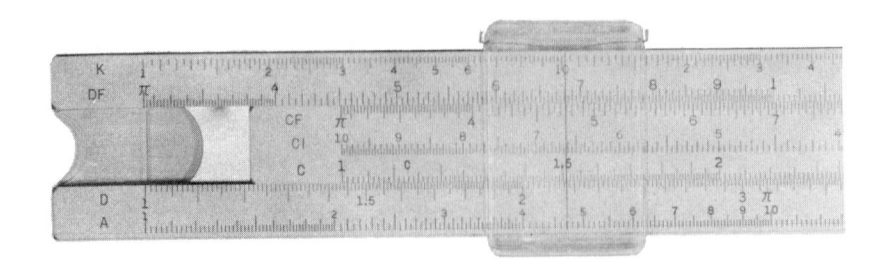

54   Quantitative Aspects of Science and Technology   [§ 4-5]

Figure 4-9.   *The product of 1.435 and 1.51.*

4-5 Division

The operation of division is the inverse of multiplication. For that reason, we should expect that division on the slide rule should be an inverse of multiplication. In multiplication, the first known factor on the D scale was placed below the appropriate C scale index. Then the indicator hairline was moved to where it crossed the C scale at the second factor. The product was found on the D scale as indicated by the hairline. These slide rule operations were based upon the principle of adding exponents for multiplication. Distances along the slide rule represent exponents, and scale marks represent corresponding numbers. In the case of division, exponents would be subtracted. Therefore, we should expect to subtract distances on the slide rule to perform divisions.

Consider the problem of dividing 6 by 3. On the slide rule we must subtract a scale distance of 1 to 3 from a scale distance of 1 to 6. The scale distance remaining would represent the exponent of the quotient, with the corresponding scale mark being the value of the quotient. Figure 4-10 shows that division. First the indicator hairline is placed over the 6 on the D scale. Then the slide is moved to where the number 3 on the C scale is placed on the hairline (without moving the indicator). Finally, the quotient is read on the D scale below the C scale index. In this case that quotient is 2, as expected.

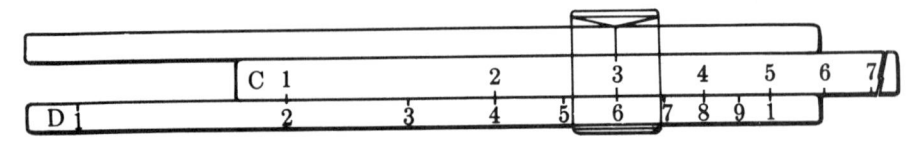

Figure 4-10.   *Six divided by three.*

[§ 4-5] Division 55

In this book we use the fraction bar to indicate division, instead of the symbol ÷. To talk about division, we must use certain words for the numbers involved. To show 12 divided by 4, we would write 12/4 = 3. The number 12 is called the *dividend*, the number 4 is called the *divisor*, and the number 3 is called the *quotient*. Because we express divisions as fractions, the numerator is the dividend, the denominator is the divisor, and the result of this division is the quotient.

Using the words discussed above to describe the process of division on the slide rule, the following steps are needed: 1, move the indicator to where the hairline is over the dividend on the D scale; 2, move the slide to where the divisor on the C scale is below the hairline, being careful not to move the indicator; 3, find the quotient on the D scale directly below the C scale index. It will be the right or left C scale index, depending upon the values of dividend and divisor selected.

Find the quotient of 28,000 divided by 0.0001268. Expressed as a fraction, this quotient would be (28,000)/(0.0001268) = ?. Because the slide rule cannot locate decimals, we must express these numbers in power of ten form before dividing. The number 28,000 would become 2.8×10^4, and 0.0001268 would become 1.268×10^{-4}. The quotient would appear as

$$(2.8 \times 10^4)/(1.268 \times 10^{-4}) = ?$$

The powers of ten can be divided first, giving

$$10^4/10^{-4} = 10^{4-(-4)} = 10^{4+4} = 10^8$$

We must be careful with this kind of division, where there is a negative exponent in the denominator, since we often forget the rule for subtraction of signed numbers (change the sign of the subtrahend and add). When the power of ten of the quotient is known, the division of 2.8 by 1.268 can be performed on the slide rule. As shown in Figure 4-11, the indicator is moved to where the hairline crosses the D scale at the dividend, 2.8. Then the slide is moved to where the hairline crosses the

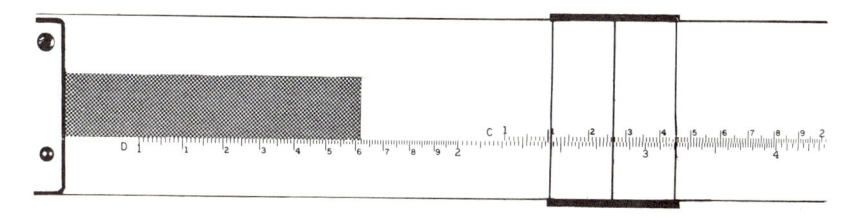

Figure 4-11. *The quotient of 2.8 and 1.268.*

C scale at the divisor, 1.268. The quotient is then on the D scale below the left C scale index. That quotient is, from Figure 4-11, 2.21. Our final answer must be 2.21×10^8.

What would 0.00003263 divided by 2300 be? The dividend is 0.00003263, and the divisor is 2300. Expressing the numbers in power of ten form, we have

$$(3.263 \times 10^{-5})/(2.3 \times 10^3) = ?$$

Dividing the powers of ten first,

$$10^{-5}/10^3 = 10^{-5-3} = 10^{-8}$$

for the power of ten of the answer. Using the slide rule we can find 3.263/2.3. Placing the hairline over 3.263 on the D scale, as shown in Figure 4-12, and moving the slide to where the hairline crosses the C scale at 2.3, we can read the quotient on the D scale below the left C scale index. That reading is 1.419. Thus our quotient is 1.419×10^{-8}.

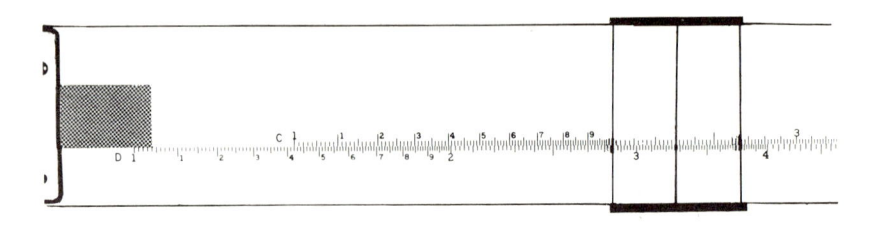

Figure 4-12.   *The quotient of 3.263 and 2.3.*

In both examples worked above, when the numbers were expressed in power of ten form, the dividend was larger than the divisor. This fact meant that the quotient of those numbers would be a number between 1 and 10, 2.21 in the first example, and 1.419 in the second. When that quotient is a number between 1 and 10, the result will always appear on the D scale below the left index on the C scale. But there are quotients for which the dividend is smaller than the divisor. For those quotients, the result will be a number between 0.1 and 1, rather than 1 and 10, and the quotient will appear on the D scale under the right C scale index instead of the left.

Using the slide rule, we divide 424,000 by 76,000,000,000. Expressing these numbers in power of ten form, we have

$$(4.24 \times 10^5)/(7.6 \times 10^{10}) = ?$$

Dividing the powers of ten, we have

$$10^5/10^{10} = 10^{5-10} = 10^{-5}$$

Then moving the indicator to where the hairline crosses the D scale at 4.24, and moving the slide to where that hairline crosses the C scale at 7.6, we read the quotient on the D scale below the right C scale index. The quotient is 0.558. The final result would be 0.558×10^{-5}. Because we normally express numbers between 1 and 10 times a power of ten, we would factor 0.558 to 5.58×10^{-1}. Then the final result would be

$$5.58 \times 10^{-1} \times 10^{-5} = 5.58 \times 10^{-6}$$

This would normally be the form in which an answer would be stated.

PROBLEMS

1.   a. 99,300/4030   b. 88,000/25,000,000   c. 24,300/582
     d. 315/808,000   e. 448,000/0.0031   f. 0.000083/44,000
     g. 0.000482/0.00315   h. 0.0029/0.0000583

4-6 *Multiplications and Divisions*

In many problems there is more involved than simply multiplying or dividing two numbers. Very often, several numbers must be multiplied and divided within the same problem. These multiple operations can be carried out on the slide rule, but the use of mental arithmetic and powers of ten then becomes essential.

As an example, consider the following problem:

$$\frac{(0.00038)(420,000)(0.0069)}{(28,000)(0.00096)} = ?$$

Before using the slide rule, one must express each number in power of ten form. Then he should perform an approximate mental calculation to get some idea what answer will result. The reason for this calculation is to locate the decimal point in the slide rule answer. Expressing each number in power of ten form, we have

$$\frac{(3.8 \times 10^{-4})(4.2 \times 10^{5})(6.9 \times 10^{-3})}{(2.8 \times 10^{4})(9.6 \times 10^{-4})} = ?$$

Performing the operations on the powers of ten first, we have

$$\frac{(10^{-4})(10^{5})(10^{-3})}{(10^{4})(10^{-4})} = \frac{10^{-4+5-3}}{10^{4-4}} = \frac{10^{-2}}{10^{0}} = 10^{-2-0} = 10^{-2}$$

Then we have remaining the slide rule problem

$$\frac{(3.8)(4.2)(6.9)}{(2.8)(9.6)} = ?$$

Now we must perform some mental arithmetic. In the numerator, we have the product of the approximations, 4, 4, and 7, which is somewhere around a hundred. In the denominator, we have about 3 times about 10, which is approximately 30. Thus we know our final slide rule answer is going to be something around 100/30, or about 3. Actually, when the numbers are multiplied and divided on the slide rule, the result may come out anywhere from 2 to 5, but our mental computation is certainly not off by as much as a factor of ten. This is why it tells us where the decimal point must be. Now we can perform these operations on the slide rule.

The simplest way of carrying out these operations is first to divide 3.8 by 2.8, then multiply this result by 4.2, divide that result by 9.6, and finally multiply that answer by 6.9. The slide rule procedure would be as follows. Move the indicator to where the hairline crosses the D scale at 3.8, and move the slide until the hairline crosses the C scale at 2.8. The answer to this operation is on the D scale below the C scale index, but we do not need to write it down. We can continue our procedure by leaving the slide fixed but moving the indicator to where the hairline crosses the C scale at 4.2, the product appearing on the D scale below the hairline. But again, we do not need this result, so we do not write it down. Next, with the indicator fixed, the slide is moved to where the hairline crosses the C scale at 9.6, the result of this division appearing below the C scale index. Of course, we still do not need this quotient, so we do not write it down either. Our final operation is to leave the slide fixed, moving the indicator to where the hairline crosses the C scale at 6.9, reading the final result where the hairline crosses the D scale. You should follow the above procedure carefully on your own slide rule if you are to gain an understanding of the steps involved.

The result of the above slide rule computation is 4.1. In our mental arithmetic, we had guessed an answer of about 3. All that process did was tell us that when our slide rule gave 4.1, the result was 4.1 and not 0.41 or 41, which are factors of ten smaller and larger than 4.1. That is precisely the reason for doing the mental arithmetic before using the

[§ 4-7]   *Squares and Square Roots*   **59**

slide rule on such problems. The final answer to the example problem must be 4.1×10^{-2}, or, without powers of ten, 0.041.

With practice on these kinds of problems, one can acquire skill in performing such arithmetic. He will learn that other scales on the slide rule can be used to make such problems even easier, and he will find that he will select the scales according to the type of problem he has. But skill in using a slide rule comes only with practice and understanding, something which can only be learned. It cannot be taught.

PROBLEMS

1. Carry out the indicated operations in one series of slide rule operations.

   a. $\dfrac{(428{,}000)(907{,}000{,}000)}{6{,}040}$ b. $\dfrac{(7820)(0.0000074)}{249{,}000}$

   c. $\dfrac{5{,}400{,}000}{(323)(529{,}000)}$ d. $\dfrac{0.0000048}{(423)(55{,}000)(0.0043)}$

2. Carry out the indicated operations in one series of slide rule operations.

   a. $\dfrac{(0.0000046)(55{,}000{,}000)(0.00000000104)}{(543{,}000)(0.00000000058)(414)(808)}$

   b. $\dfrac{(414{,}000)(0.0082)}{(4145)(209{,}000)(0.038)(0.144)(0.000074)}$

4-7 *Squares and Square Roots*

The problem of finding squares or square roots on the slide rule is relatively simple. It does not involve any movement of the slide, requiring only the A and D scales and the indicator. Square root problems do require that numbers be expressed as a number between 1 and 100 times an *even* power of ten. Also, in using the A scale, the location of digits depends upon whether the number is between 1 and 10 or between 10 and 100. These points will be exemplified in the following problems.

What is the square root of 6,150,000? Expressing this number between 1 and 100 times an even power of ten, we have 6.15×10^6. Using the fact that the square root of a number is the same as that number to the 1/2 power, the power of ten square root can be evaluated. That root is

$$(10^6)^{1/2} = 10^{6/2} = 10^3$$

Using the slide rule to find the square root of 6.15, the digits 615 must be found on the A scale. But there are two places where these digits occur. Which one should be selected?

An examination of the A and D scales shows that the A scale has two scales compressed into the same length as the D scale. As numbers on the A scale go from 1 on the left to the next 1 at the middle of the scale, numbers on the D scale go from 1 on the left to about 3.16. As numbers on the A scale go from the middle 1 to the 1 on the right, the D scale numbers go from 3.16 to 1 on the right. We can think of the A scale as going from 1 to 10, then from 10 to 100, whereas the D scale goes from 1 to 10. On some slide rules, the 10 is marked for the center 1 of the A scale, and the right 1 on the A scale is marked with 100. The left half of the A scale is used to find square roots of numbers between 1 and 10, and the right half of the A scale to find square roots of numbers between 10 and 100.

In the above sample problem, we are trying to find the square root of 6.15. Since this number is between 1 and 10, we would use the left half of the A scale. Moving the indicator to the digits 615 on the left half of scale A, as shown in Figure 4-13, the square root is where the hairline crosses the D scale at 2.48. The final answer to our problem would then be 2.48×10^3, or, without powers of ten, 2480.

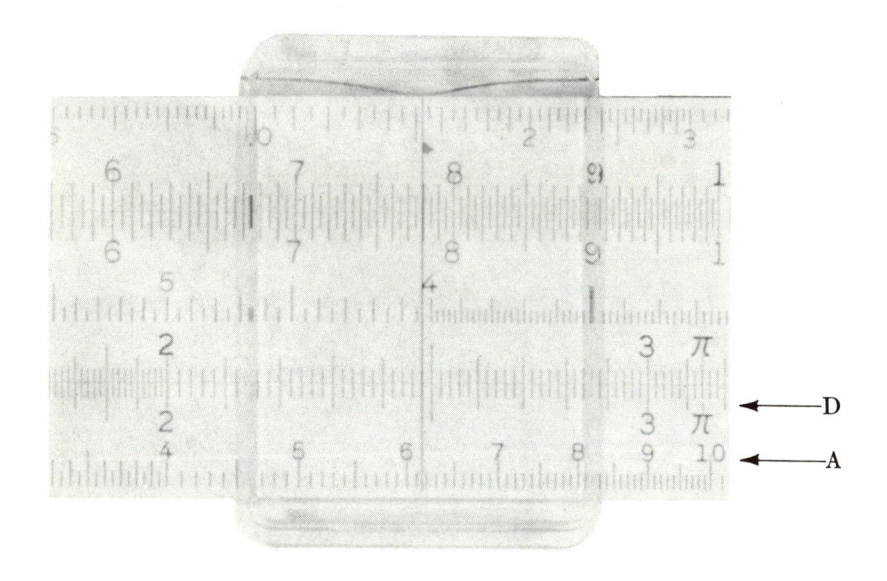

Figure 4-13. *The square root of 6.15.*

[§ 4-7]   *Squares and Square Roots*   **61**

As another example, find the square root of 0.0000005480. In this case, if the number were expressed in ordinary power of ten form, it would be 5.480×10^{-7}. Although one could find the square root of 5.480 on the slide rule, an odd power of ten, like 10^{-7}, does not have an integral power of ten as a square root. For that reason, we must express this number as 54.80×10^{-8}, a number between 1 and 100 times an even power of ten. Finding the square root of 10^{-8}, we have

$$(10^{-8})^{1/2} = 10^{-8/2} = 10^{-4}$$

Because 54.80 is between 10 and 100, the right half of the A scale must be used. As shown in Figure 4-14, moving the indicator to where the hairline crosses the A scale at the digits 5480, the square root can be read under the hairline on scale D as 7.4. The final result must be 7.4×10^{-4}, or, written without powers of ten, it is 0.00074.

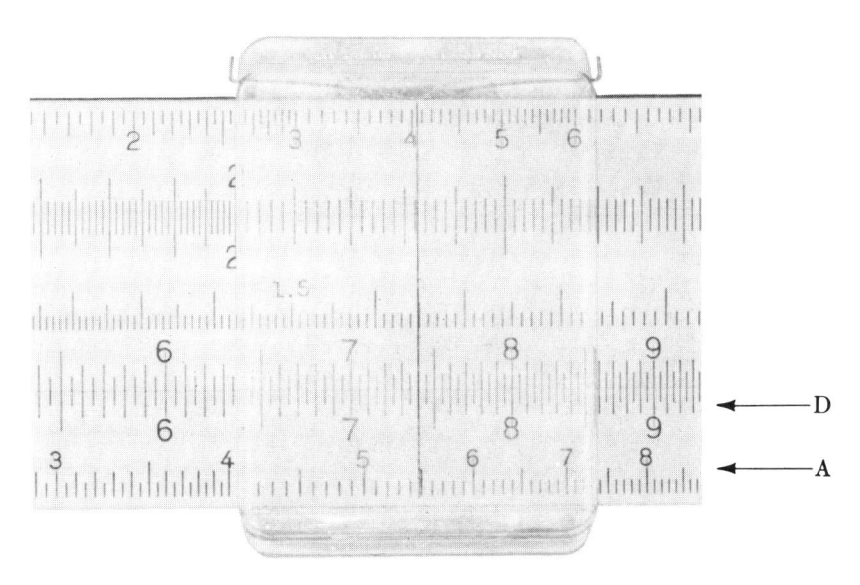

Figure 4-14.   *The square root of 54.8.*

As a last example, we shall find the square root of the number 74,000. Writing 74,000 as a number from 1 to 100 times an even power of ten, we have 7.4×10^4. Taking the square root of 10^4,

$$(10^4)^{1/2} = 10^{4/2} = 10^2$$

As shown in Figure 4-15, moving the indicator to where the hairline crosses the left half of the A scale at the digits 740, the square root is

62   *Quantitative Aspects of Science and Technology*   [§4-7]

read under the hairline on the D scale as 2.72. The square root of 74,000 is therefore 2.72 × 10², or, without powers of ten, simply 272.

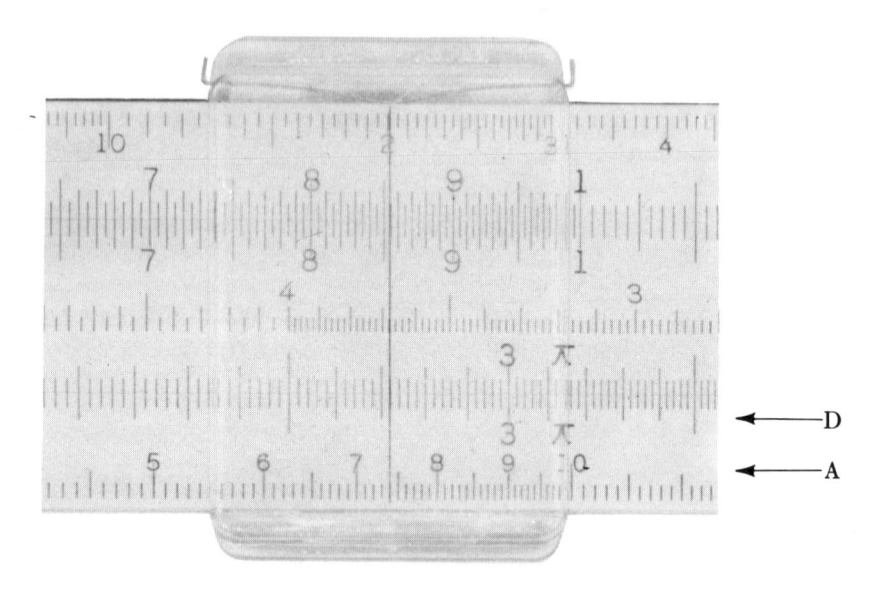

Figure 4-15.   *The square root of 7.4.*

The process of squaring numbers can also be accomplished using the A and D scales of the slide rule. Because this process is the inverse of finding square roots, the procedure is immediately apparent. Locate the number to be squared on the D scale by moving the indicator hairline to cover that number. Then read the square under the hairline on the A scale. It is still necessary to use powers of ten, but, when squaring numbers, they should again be put into the form of a number between 1 and 10 times a power of ten.

As an example, we might square the number 0.0074. Expressing it in power of ten form, we have 7.4×10^{-3}. Squaring the power of ten, we get

$$(10^{-3})^2 = 10^{(-3)(2)} = 10^{-6}$$

Here we have used the algebraic rule, $(10^m)^n = 10^{(m)(n)}$, mentioned in the previous chapter. From Figure 4-14, moving the indicator to the digits 74 on the D scale, we find the square on the A scale under the hairline as digits 548. Since we were squaring the number 7.4, we know that the result is the number 54.8. The answer must be 54.8×10^{-6}. If we want to get this number into the correct power of ten form, we can factor 54.8 into 5.48×10^1. Then we have

$$5.48 \times 10^1 \times 10^{-6} = 5.48 \times 10^{1-6} = 5.48 \times 10^{-5}$$

for the final answer.

This chapter has been concerned with the operations of multiplication, division, and finding squares and square roots on the slide rule. It has been made clear that the slide rule alone cannot perform these computations. The reader must understand and be able to use powers of ten. He must be able to do mental arithmetic to make some computations possible. Yet with practice, the slide rule can become an invaluable aid to the student who wants to learn science or technology. Its use can free him from laborious pencil and paper calculations so that he can concentrate on learning the basic principles of those subjects. Of course there are other scales, not discussed here, on most slide rules. It is expected that when the reader has gained skill and confidence in the use of his own slide rule, he will learn to use other scales as the need arises.

PROBLEMS

1. Find the squares of the following numbers using the slide rule.
   a. 345,000,000,000   b. 0.0000000472   c. 5.44×10^{-33} d. 9.2×10^{15}

2. Find the square roots of the following numbers using the slide rule.
   a. 974,000,000,000   b. 8343   c. 0.443   d. 0.0488
   e. 0.00327   f. 0.00804   g. 0.00066   h. 0.00000000000757
   i. 0.00000000000000482   j. 3.32×10^{23} k. 4.92×10^{-15}
   l. 493,000,000,000   m. 98,000   n. 8.84×10^{-34}

5 Solving Algebraic Equations

5-1 Algebraic Symbols

Although a scientist or technician does spend much time working with numerical symbols in the rational numbers, he very often needs to use symbols to represent physical quantities which are unknown. He may know the quantity is of a certain kind, but not know its value. To represent these quantities he uses letters of our alphabet, and of others, like the Greek alphabet. These symbols are used to stand for real numbers, although at times they are also used to stand for complex numbers.

For example, an engineer may want to represent the distance from the earth's surface to some unspecified point in space. Because he does not know what specific distance is to be used, he lets the letter H, h, d, R, or perhaps X, stand for that distance. The symbol selected is not important, as long as what the symbol represents is clear. Most often, symbols are letters related to one of the words in the quantity, like the first letter. For example, F is used to represent force, P to represent power, and W to represent work. But P is also used to represent pressure. How do we know which is meant? When a symbol is used, its meaning must always be explained.

Alphabetic symbols are used by scientists and engineers in several different ways. In the preceding paragraph, symbols used as "variables" were discussed. They represent physical quantities which have not yet been specified. Perhaps the value of a quantity would depend upon several other variables. A number of other quantities may be given; then by some algebraic formula or process the "unknown variable" can be determined. Another way symbols are used is in representing what are called *constants*. In science and technology there are many physical measurements which, once made, do not change. These numbers are known, therefore it would not seem necessary to use symbols to represent them. But by using symbols instead of the actual numbers, it is easier to write equations involving the quantities; also, there are some constants that are unchanging on earth, but may change out in space, or on some other planet. If the law involving these quantities is to be correct, the use of a symbol instead of the measured value on earth enables the scientist to write a more general law. As an example, the letter g is used for the acceleration of gravity instead of the measurement 9.8 m/sec². The value of g is almost constant over the earth's surface, but it changes radically as one moves away from the earth.

Other constants are universal; they do not change no matter where they are determined. The constant of universal gravitation,

$$6.67 \times 10^{-11} n\text{-}m^2/\text{kg}^2$$

is represented by the letter G. All available evidence shows that this constant is the same everywhere within the visible universe. The only reason for using G instead of its numerical value is that G is easier to write, and the numerical value of the constant depends upon the units selected. When computations are to be made, the engineer or scientist will "look up" the correct value of G. There are not many universal constants, but another is Planck's constant, $6.6 \times 10^{-34} J\text{-sec}$. This constant is represented by the letter h. It is a fundamental constant of modern atomic physics. The letter h is used in equations of modern physics until the scientist must make a computation. Then he replaces the h by this number.

We use the words *variable*, *unknown*, and *constant* to refer to the alphabetic symbols representing physical quantities. The only real constants are the universal constants, and even these are expressed in a certain system of measurement.

In his study of algebra, the student almost always uses particular letters of the alphabet. Normally, such courses use letters near the beginning of the alphabet for constants, i.e., *a,b,c,d*, and letters near the end of the alphabet for unknowns, i.e., *u, v, w, x, y, z*. Unfortunately, people become so accustomed to using these letters in algebraic processes that when they study science, where all sorts of letters and even some Greek letters are used, they are confused for absolutely no reason. If the algebra is understood, it should not make any difference what symbols are used. As an example, most people could solve the equation

$$X^2 + 4X + 4 = 0$$

But if the symbol ψ is used instead of X, they appear helpless in solving the equation. They say "What does that ψ mean? I can't solve the equation until I know what it is." The equation would be,

$$\psi^2 + 4\psi + 4 = 0$$

They know as much about this Greek symbol, ψ, as they do about X; it is just that they are unfamiliar with using such symbols. Many different kinds of symbols are used in science and technology, and one must adjust to their use, remembering that any symbol can be used to represent a physical quantity.

PROBLEMS

1. Select a letter symbol to represent each of the following phrases.
   a. the speed of the airplane   b. the temperature of the gas
   c. the volume of the container   d. a man's age

2. In atomic physics, the frequency of radiation f is given by the formula $f = E/h$, where E is the energy of the radiation and h is Planck's constant. If $h = 6.6 \times 10^{-34}$ J-sec, what is the energy of radiation of frequency $f = 4 \times 10^{10}$ sec^{-1}?

3. In each of the following equations, change the variable from X to the variable shown.
   a. $X^2 \times 3X - 5 = 0$ ψ b. $(1/2)Xv^2 + Xgh + P = C$ ρ
   c. $X^3 - 8 = 0$ β

5-2 Changing Verbal Statements to Symbolic Form

One of the greatest difficulties in quantitative courses is analyzing verbal statements. The scientist or engineer has a similar, but considerably more difficult problem, in analyzing a physical situation. He must convert the physical situation into verbal statements which can be put into quantitative form. The kind of ability needed to do these analyses must be learned through effort. One of the first skills to be acquired is the ability to change verbal statements into symbolic form.

If verbal statements can be changed to symbolic form, then algebra must have a grammar and a punctuation. Certain symbols must represent certain words or groups of words in an exact way. Let us examine some of those symbols and the words they represent.

The symbol = means *is*, or *is equivalent to*, or *has the same value as*. One is tempted to say that the symbol = means *is the same as*, but this statement is incorrect, since, for example, 5 = 3 + 2 is true, but 5 is not the same as the number 3 next to a plus next to a 2. The number 5 is equivalent to 3 + 2, and it has that value.

The symbol + means that the quantity to the right of + is a positive number, or, if placed between two numbers, it means to perform the algebraic process of addition on the numbers. Since it is usually used in the addition process, it could replace the words *find the sum of two numbers*.

The symbol − means that the quantity to the right of − is a negative number, or, if placed between two numbers, it means to perform the algebraic process of subtraction on the numbers. When it is used in subtraction, it could replace the words *find the difference of two numbers*.

The process of multiplication requires the use of a dot between products, or an ×, if the × is not confused with an unknown. Sometimes sets of parentheses are used to indicate multiplication. When letter symbols are used, placing the letters next to each other with no marks of any kind means that they are to be multiplied. Thus, RT, means to multiply R and T when the numbers they represent are known. Because the multiplication cannot be carried out until the values are known, this representation is called an *indicated multiplication*. In the process of division, we use as a symbol the slant bar /, placing the dividend above it, and the divisor below it. The symbol is the same whether we use letters or numbers.

Let us consider certain examples of verbal statements. For the first example, we will change a sentence into symbolic form. The sentence is "The sum of two numbers is 43." To analyze this verbal statement, we must first read it. Next, we notice that the words "The sum of two numbers" must have the same value as the number 43, because of the use of the word "is". We can now use an equal sign to separate that phrase from the number 43. "The sum of two numbers = 43." Next we observe that there are two unknown numbers. We must name them, and we must do it explicitly. Let us call one of them R, and the other T. If we had more information about them, we might be able to call one a larger number, and the other smaller. But we have no such information yet. Our statement now becomes "The sum of R and T = 43." We have left the problem of changing the words "The sum of" to symbolic form, but we know that these words mean the operation of addition. The plus sign indicates that process. The final algebraic statement becomes $R + T = 43$.

Consider another verbal statement. "A man's age is now five ninths of what it will be 28 years from now." Again, the word "is" tells us where an equal sign should go. Thus we have "A man's age now = five ninths of what it will be 28 years from now." We can replace "A man's age now" by some symbol, say N. Then the statement becomes "N = five ninths of what it will be 28 years from now." In reading the remaining statement, we recognize that the words "what it will be 28 years from now" mean the man's age in 28 years. Those words can be replaced by "$N + 28$", therefore the statement becomes "N = five ninths of $(N + 28)$." The parentheses are used because it is five ninths of the age later, not the age now. The parentheses group the symbols $N + 28$ into one quantity. Finally, we must symbolize the words "five ninths of". But these words mean to multiply by 5/9. We have then the final algebraic statement "$N = (5/9)(N + 28)$."

We could consider many examples of the type shown above, but one must learn this process of analysis by actually doing it. Certain specific phrases occur quite frequently in verbal problems. Some of those phrases are listed on page 69, with the corresponding symbolic form. Study them carefully.

These are but a few of the kinds of verbal statements which can be changed to symbolic form. The basic method of analysis is simple. Read the entire verbal statement. Decide what unknown quantities are present; then specify what symbols are to represent those quantities. Finally, look for words like "is", "times", "more than", "less than", "increased by a factor of", or "decreased by a factor of", replacing them by the appropriate symbol, =, +, ×, or /. With experience, one can change verbal statements to symbolic form with surprising ease. The panic we experience in first reading a complicated passage for conversion to symbolic form disappears as soon as we concentrate on converting the passage in small parts at a time.

[§ 5-2] Changing Verbal Statements to Symbolic Form   69

Verbal statement	Symbolic form
Their difference is	$U - V =$
Five times the smaller	$5R$
One seventh of his age 10 years hence	$(1/7)(A + 10)$
How long after the express train leaves	T
A bus increases its speed by 10 mph	$V + 10$
A man walks a distance in 40 min less than his son	D in $T - 40$
Depreciates in value to 20 per cent of the original value	$0.20V$
If the radius of the earth's orbit were doubled	$2R$
The period of revolution of the moon is 27.3 days	$T = 27.3$
The height above sea level	h
How far does the body move	$D = ?$
Five less than the other number	$N - 5$
Seven greater than the number	$M + 7$

PROBLEMS

1. In each of the following equations, Y stands for "the vertical displacement", T stands for "the time required", g stands for "the acceleration of gravity", V stands for "the velocity at time T", and V_0 stands for "the initial velocity". Change each of the following equations into complete verbal statements using the quoted phrases above and the correct meaning for operations indicated.
   a. $Y = V_0 T - (1/2)gT^2$ b. $V = V_0 - gT$ c. $V^2 = V_0^2 - 2gY$

2. In each of the following expressions, X stands for the number 6 and Y stands for the number 3. Carry out what the symbols tell you must be done to get a numerical value for the expression.
   a. XY b. $X/6$ c. X/Y d. X^2Y^3 e. $(X^2 + 3Y^4 - XY^2)/X$

3. Convert each of the following verbal statements or phrases into symbolic form, carefully defining any unknowns selected.
   a. An airplane has a speed which is
   b. The sum of the squares of two numbers is 20
   c. If the price per item were decreased by 3 per cent
   d. One number is 6 greater than 3 times another number
   e. One man's age is 12 years less than 3 times his son's age

5-3 Mathematical Operators

In working with physical quantities, the scientist or technician must perform a number of different kinds of *operations*. You are already familiar with several operators. An *operator* is a symbol which indicates what to do with one or a pair of numbers. The plus sign, when used to indicate addition, is an operator. It is placed between two numbers to indicate that the two numbers are to be added. The multiplication symbol, either a dot or a cross, acts as an operator when placed between two numbers, indicating that the two factors are to be multiplied. The slant bar placed between two numbers is an operator which indicates that a division is to be performed on the numbers involved.

There are many different operators which act on single numbers. Although one may not be familiar with the term, he has used operators. For example, the symbol $\sqrt{}$ is an operator. It replaces the words "take the square root of the number which follows". Of course, the words "square root" must be defined. Using this operator, $\sqrt{4}$, indicates that we are to take the square root of 4. We are accustomed to seeing a bar over the top of the number, but it is not necessary unless there is more than one term involved, since it is only a grouping symbol. As an example, the bar is needed in $\sqrt{N + M}$ if we want to take the square root of the entire sum and not just the first term, N. But that operation could also be indicated by $\sqrt{(N + M)}$, where the parentheses have been used for grouping instead of a bar.

In advanced mathematics, other operators are used. One such operator is $\int () dx$. It indicates that *the integral of the quantity in parentheses must be determined*. Most beginning college students have never seen this operator, but engineers and scientists use it quite frequently. They also use many other operators, and in modern physics much of the theory is developed in terms of operators without even specifying exactly what the operator must be. For certain kinds of operators, important results come out of mathematical analyses using them. When these results are specified to definite operations, the scientist has learned something about atomic physics.

PROBLEMS

1. We shall invent some operators. Let us decide that a circle placed around a quantity means to take the reciprocal of that quantity. For example,

$$\textcircled{4} = 1/4 = 0.25$$

[§ 5-4] Mathematical Equations 71

Let us invent another operator, say a triangle placed around a quantity. Let the triangle operator mean to square the quantity and subtract 3 from that result. For example,

$$\triangle{5} = 5^2 - 3 = 25 - 3 = 22$$

Using these two operators, evaluate each of the following expressions.

a. $\triangle{6}$ b. $\bigcirc{6}$ c. $\triangle{4} - \bigcirc{3}$ d. $\triangle{5}\bigcirc{7}$ e. $\bigcirc{8}\bigcirc{2} - \bigcirc{6} / \triangle{3}$

2. The operator $\sqrt{}$ means to find the square root of the number which follows. Evaluate each of the following quantities.

a. $\sqrt{9}$ b. $\sqrt{9} + \sqrt{16}$ c. $\sqrt{(9 + 16)}$ d. $\sqrt{9} \times \sqrt{4}$ e. $\sqrt{(9 \times 4)}$

5-4 *Mathematical Equations*

Science and engineering problems almost invariably involve, at some point, the task of "solving" an equation. When a physical situation is put into symbolic form, the result is an equation or set of equations that must be solved. The processes involved in solving these equations are what we learn in college mathematics courses.

What is an equation? This question must be answered before the problem of solving it is considered. When an equation is written, it contains three parts: the left side or member, the equal sign, and the right side or member. The left and right members (or sides) of an equation can be interchanged without changing anything, since the equal sign means "is equivalent to". Any combination of symbols and operations can be included on either side of an equation, but the equal sign should not be used unless the left side is in fact equivalent to the right side. There are two types of equations. One type is called a *conditional equation*, the other an *identical equation*, or *identity*. A conditional equation is a statement which is true only for a certain value of the unknown. For example, the equation $X - 3 = 7$ is true only when $X = 10$. It is therefore a conditional equation. An identical equation is true for *all* values of the unknown. For example, the equation $X - 6 = X - 6$ is true for any number X. It is an identical equation. Many identities occur in trigonometry, where some equations are true for any angle.

An equation therefore consists of three parts, two of which are equivalent to each other, the third part being a symbol indicating this equivalence. Because of this definition of an equation, *any* operation performed on one member of an equation will give exactly the same result as performing the same operation on the other member. One would be performing an identical operation on equivalent quantities;

therefore, the results must be equivalent. All of this talk means that *any* operation can be performed on one side of an equation, if exactly the same operation is performed on the other side of that equation. The entire mathematics curriculum of the engineering or science student is devoted to learning those operations and performing them on certain kinds of equations. A fundamental principle of algebra can therefore be stated as follows. Anything (any mathematical operation) can be done (performed on) to one side of an equation, if exactly the same thing is done to the other side of that equation, without changing the equivalence.

PROBLEMS

1. Which of the following equations are conditional equations? Which are identical equations?
   a. $X = 6$ b. $X = X$ c. $(X - 2)^2 = X^2 - 4X + 4$
   d. $X - 7 = 3$

2. Defining the triangle operation as squaring a number and subtracting 5 from the result, for example, $\triangle_3 = 9 - 5 = 4$, perform the triangle operation to both sides of each of the following equations.
   a. $X = 2$ b. $Y = 2X^2$ c. $Y/X = 5$

5-5 Solving Algebraic Equations

The basic problem in algebra is to "solve" an equation for a particular quantity. The word *solve* means simply to get that particular quantity by itself on one side of an equation. The ways in which that quantity reaches its lonely position on one side of an equation involve whatever operations the student can imagine which will lead to that result. But it is essential that the operations be performed on *both* sides of the equation. Of course, it is not just a haphazard selection of operations that will solve an equation. There are good reasons for selecting operations to be performed on both sides of an equation.

Let us consider a few typical examples, and how they are solved. We shall use X, Y, and Z for unknowns, and A, B, and C for constants. The equation $X + A = B$ is to be solved for X. Solving the equation means getting X by itself on one side of the equation. We are free to perform any operation on one side, provided that we do the same thing to the other side. We might just try some operation, but the equation itself gives us direction. The unknown will always occur with operations being performed on it. If we want to solve the equation, we perform the *inverse* of indicated operations to both sides of the equation. In this equation, A is added to X. Thus, if we want to solve for X, we must perform the inverse of addition of A to both sides of the equation. Subtraction is the inverse of addition, therefore we must subtract A from both sides of this equation. Performing this operation on both sides of the equation, we have

$$X + A - A = B - A$$

and our solution to the original equation is

$$X = B - A$$

We have solved the equation $X + A = B$ in terms of B and A.

We might be given a problem of the type $AX = B$. In this equation, X is multiplied by A. The inverse of multiplication is division; therefore, to solve for X we must perform the operation of division by A to both sides of this equation. Performing that operation, we have

$$(A)(1/A)X = (B)(1/A)$$

or, since $(A)(1/A) = 1$,

$$X = B/A$$

How would the equation $X - A = B$ be solved for X? In this case, the indicated operation on X is subtraction of A. What operation should be performed on both sides of the equation? We perform the inverse of the indicated operation; therefore, we must add A to both sides of this equation. Adding A to both sides, we have

$$X - A + A = B + A$$

or, since $-A + A = 0$, the solution is

$$X = B + A$$

How would the equation $X/A = B$ be solved for X? In this equation, the indicated operation on X is division by A. What operation must be performed on both sides of this equation? The inverse of division is multiplication; therefore we must multiply both sides of the equation by A. Performing that operation, we have

$$X(1/A)(A) = (B)(A)$$

or, because $(1/A)(A) = 1$, the result is

$$X = BA$$

Sometimes more than one operation involving the unknown is indicated. For example, we might have an equation like $AX + B = C$, or $AX - B = C$. But if the indicated operations on X were to be performed, they would have to be done in a certain order. We would have to multiply before adding or subtracting. For that reason, if we are going to perform inverse operations on both sides of the equation, we should do so in reverse order. The operations are indicated first as multiplication, then addition or subtraction. The inverse operations to be performed on both sides of the equations would be in the order: subtraction or addition; then division.

Let us consider the equation $AX + B = C$. In this equation, if we knew the values of A, X, and B, we would perform multiplication of A and X first. Then we would add the result to B to get a quantity which would be equal to C. If we want to perform the correct set of operations on both sides of this equation to solve it, we should first subtract B from both sides of the equation; then we should divide both sides of the equation by A. Performing the subtraction of B first, the result is $AX = C - B$. Then performing the operation of division by A on both sides of this equation we have

$$X = C/A - B/A$$

The original equation is then solved for X. If the values of A, B, and C were known, the symbols could be replaced by those numbers, and the indicated operations carried out to give the numerical value of X.

How would the equation $X/A - B = C$ be solved for X? In this problem the operation of division of X by A is indicated first; then the operation of subtraction of B is indicated. If this equation is to be solved, the inverse of these operations must be performed on both sides of the equation in reverse order. The inverse of subtraction is addition, therefore B must be added to both sides of this equation as a first step. Then, because multiplication is the inverse of division, A must be multiplied by both sides of the result of the last operation. Performing the addition of B first, the equation becomes $X/A = C + B$. Next, multiplying both sides of that result by A, the solution is $X = CA + BA$. Again, if the values of A, B, and C were known, the symbols could be replaced by those numbers and the indicated operations carried out to give a numerical value for X.

Most often, equations in science and engineering are more complicated than the ones just considered. But the basic procedure for solving them is the same. First examine the equation to determine the order in which the indicated operations would be performed on the unknown. Next, list those steps in reverse order. Finally, perform the inverse of each operation in the list, one at a time, to *both sides* of the equation. If this procedure is followed carefully, the unknown will finally be by itself on one side of the equation, and it will be solved.

As a final example, consider the following rather complicated-appearing equation, which we want to solve for X.

$$\frac{BX^2 + A}{C - D} + E = F$$

where A, B, C, D, E, and F, are constants whose values are known. To say that we want to solve the equation for X means to get X by itself on one side of this equation. Let us first make a list of the indicated operations involving X, in the order that they would have to be performed.
 1. Square X.
 2. Multiply the result of 1. by B.
 3. Add A to the result of 2.
 4. Divide the result of 3. by the quantity $(C - D)$.
 5. Add the result of 4. to E.

Let us now reverse the order of these operations and change each operation to its inverse. Remember that the inverse of adding is subtracting; the inverse of multiplying is dividing; and the inverse of squaring is taking the square root. The list would be as follows.
 1. Subtract E from both sides of the equation.
 2. Multiply both sides of the equation by the quantity $(C - D)$.
 3. Subtract A from both sides of the equation.
 4. Divide both sides of the equation by B.
 5. Take the square root of both sides of the equation.

Now let us perform these operations, one step at a time. Performing 1., we have

$$\frac{BX^2 + A}{C - D} = F - E$$

Next, doing 2. gives

$$BX^2 + A = (F - E)(C - D)$$

Then, carrying out 3. gives

$$BX^2 = (F - E)(C - D) - A$$

Next, doing 4., we have

$$X^2 = [(F - E)(C - D) - A]/B$$

Finally, performing 5., we get the solution

$$X = \sqrt{[(F - E)(C - D) - A]/B}$$

If the various values of A, B, C, D, E, and F were known, those symbols could be replaced by their numerical values; then the indicated operations could be carried out to find the numerical value of X.

The preceding discussion illustrates the basic method for solving equations. Any mathematical operation can be performed on one side of an equation if the same operation is performed on the other side of the equation. To decide what operations are to be performed in solving an equation, the indicated operations on the unknown and the order in which they are to be performed must be determined. By reversing the order and performing the inverse of each of these operations to both sides of the equation, it can be solved for the unknown.

Although we have discussed methods of solving equations, and some of the operations involved, one must have a prior, basic understanding of algebra to solve equations. He must understand the use of grouping symbols, and he must know the basic rules of algebra involving exponents and radicals. He must already understand the commutative, associative, and distributive laws as learned in an algebra course. Many equations require that a process called factoring be accomplished before the steps outlined above can be carried out. Most college students have studied algebra at some time in their lives. It is hoped that the material of this chapter will fill the forgotten gaps and, perhaps, enhance his understanding of the fundamental principles of algebra to such an extent that he can recall what he has forgotten and perform the kinds of algebraic processes required in science and technology courses. If the student has a specific deficiency in algebra, he should identify it and review that topic in an elementary algebra textbook before working the problems which follow.

PROBLEMS

1. Solve each of the following equations which has a solution in the set of *natural numbers*. If the equation has no solution in the set of natural numbers, indicate the problem by the statement "no solution".
   a. $3X = 2$ b. $X - 4 = 7$ c. $X + 8 = 3$ d. $5X = 25$
   e. $X + 3 = 8$ f. $X - 2 = 0$ g. $8X = 2$ h. $X/3 = 5$
   i. $X/4 + 5 = 7$ j. $X/3 + 8 = 5$

2. Solve each of the following equations which has a solution in the set of *integers*. If the equation has no solution in the set of integers, indicate the problem by the statement "no solution".
   a. $3X = 2$ b. $X/4 - 2 = 7$ c. $-8X = 32$ d. $-X - 4 = -88$
   e. $3/X - 2 = 1$ f. $X/3 = -5$ g. $7X = 5$ h. $X + 5 = 3$

[§ 5-5]   *Solving Algebraic Equations*   77

3. Solve each of the following equations which has a solution in the set of *rational numbers*. If the equation has no solution in the set of rational numbers, indicate the problem by the statement "no solution".
   a. $X/3 = 4.2$ b. $5/X = -8$ c. $X = -3$ d. $5X/3 - 4.2 = 5.5$
   e. $-4X = 16$ f. $4/X + 1/3 = 1/5$ g. $X^2 - 4 = 0$
   h. $X^2 + 4 = 0$

4. By performing the indicated operations in the correct order, evaluate each of the following expressions. Let $A = -2$, $B = -3$, $C = 1$, $D = 2$, $E = 3$.
   a. $\dfrac{3A - B}{C + D}$ b. $\dfrac{AB - C/D}{A^2 - B^2}$ c. $B^2 - C^2$ d. $\dfrac{-BE}{C} - D^2 + A$
   e. $\dfrac{ABCDE}{C + D} + \dfrac{1}{ABC}$

5. Write the order of the indicated operations on the unknown X in the following equations.
   a. $\dfrac{3X^2 - 4}{2} + 5 = 6$ b. $\dfrac{AX^3 - B}{C + D} - E = F$ c. $\dfrac{A(X - B)}{C - D} + E = F$

6. The following equations are laws or definitions taken from science and engineering. By listing the indicated operations on the unknowns, then reversing that order and performing the inverse operation on both sides of each equation, solve for the unknowns.
   a. $F = ma$ for a
   b. $a = \dfrac{V_2 - V_1}{t}$ for V_2
   c. $V^2 = V_0^2 + 2ax$ for x
   d. $T = \sqrt{2h/g}$ for h
   e. $(M - ut)a = uv$ for M
   f. $E = (1/2) MV^2$ for V
   g. $\omega^2 = \dfrac{2uh}{mr^2 + J}$ for h
   h. $E = (1/2) MV^2 + (1/2) IW^2$ for V
   i. $E = mgR(1 - R/r)$ for g

6 Introduction to Ratio and Proportion

One of the topics least understood initially, yet frequently used in science, engineering, and technology, involves ratio and proportion. To many people, the way a table of numbers is analyzed to form an equation seems mysterious. The general way in which proportions are discussed often leaves us with the feeling that this subject must remain remote and incomprehensible. But the concepts of ratio and proportion are fundamental to an understanding of the principles of science and technology. There are clear definitions and procedures for using ratios and proportions; therefore, one should master these techniques as early as possible.

6-1 The Meaning of Ratio

It is very often desired to express how much larger one quantity is than another. For example, if a room is twice as long as it is wide, some short, mathematical way of stating this fact is needed. Frequently, the comparison of two quantities must be made. Sometimes these quantities are of the same kind, but sometimes they are not. If two quantities are to be compared, it could be simply a matter of deciding which of the two is larger. This comparison might be made by finding the difference of the two quantities, but such a comparison would have the disadvantage of depending upon the particular unit of measurement used. Also, if the quantities were not of the same kind, it would not be possible to find their difference.

As an example, two students measure the length and width of a room. One student uses a meter stick, the other uses a foot ruler. The measurements are, in metric units, length 4.312 m, and width 3.815 m. In English units, the measurements are length, 14.13 ft, and width 12.52 ft. To compare the length and the width, one *could* simply subtract the smaller dimension from the larger dimension, getting 0.497 m, or 1.61 ft. But when the comparison is made in this way, the results depend upon what system of units is used. It would certainly be better if comparisons could be made between the same kinds of quantities in such a way that the unit of measurement would not affect the result.

There are other disadvantages in using differences for comparing two quantities. For very large values of the quantity, a certain difference might be quite insignificant, yet that same difference for very small values could be significant. For example, a large piece of farmland might have a length of 4,128.6 ft and a width of 4,128.1 ft. The difference is 0.5 ft. But a book might have a length of 0.7 ft and a width of 0.4 ft, for a difference of only 0.3 ft. As a comparison, the book is much longer than it is wide, but the farmland is nearly square. Differences do not reveal these facts.

It is clear that some method, other than simply finding differences, is needed for making quantitative comparisons. What is needed is some comparison which will tell how much larger one quantity is than another. That comparison can be made by dividing one quantity by the other.

Using the examples just mentioned, the length of the farm divided by its width is

$$4128.6 \text{ ft}/4128.1 \text{ ft} = 1.000121$$

Notice that feet divided by feet "cancels" out, so that the result, 1.000121, is a *pure* number without units. The same method indicates that the length of the book is 1.75 times longer than its width.

Using divisions as comparisons has another advantage. Consider the example involving the measurement of the length and width of a room in metric and English units. When the length is divided by the width in metric units, we get 4.312 m/3.815 m = 1.13. When the length is divided by the width in English units, we get 14.13 ft/12.52 ft = 1.13. The same result was found, regardless of the units used. This is an important fact. When measurements of the same kind are divided, the result is a pure number and is completely independent of the units used. If the length and width had been measured in "bleeps" the length divided by the width would still have been 1.13. The length of this room is 1.13 times longer than its width.

This process of comparing two quantities by dividing one by the other is called finding a *ratio* of one quantity to the other. When we divided the length of the room by its width, we were finding the ratio of the length to the width. Whenever a ratio of two quantities is found, say the ratio of X to Y, written X/Y, the quantity X is divided by Y to give some ratio R, stated algebraically as $X/Y = R$. The meaning of the ratio is then simple. In the ratio $X/Y = R$, read "the ratio of X to Y is R", we are stating that X is R times larger than Y.

When Johannes Kepler (1571-1630) created his laws of planetary motion out of the vast data collected by Tycho Brahe (1546-1601), Kepler had no measurements of the actual distances from the planets to the sun. He had to use ratios of the other planet-sun distances to the earth-sun distance. Table 6-1 shows those ratios.

TABLE 6-1
The ratio of planet-sun distances to the earth-sun distance

Planet	Ratio
Mercury	0.387
Venus	0.723
Earth	1.000
Mars	1.524
Jupiter	5.21
Saturn	9.53

By the definition of the term *ratio* given above, one could state that the planet Mercury is 0.387 times farther away from the sun than is the earth. This number is, of course, less than 1; therefore, we know that

Mercury is only slightly more than 1/3 as far from the sun as is our earth. Jupiter must be 5.21 times farther away from the sun than is the earth. The significant point about these ratios is that the actual distance measurements could be made in *any* unit, but the ratios would remain the same.

Sometimes ratios of quantities of a different kind are used. The reason for their use is that it is often possible to make a measurement of a ratio, but not of the two quantities involved in the ratio. When and if either of these two quantities can be measured, then the other one will be known from the ratio measurement.

As an example, in the development of our knowledge about atomic particles, the ratio of electric charge to the mass of the electron could be determined in one experiment. But the experiment could not tell either the charge or the mass of the electron. In another different kind of experiment the electric charge of the electron could be measured. When the charge to mass experiment was performed, the result was

$$e/m = 1.758 \times 10^{11} \text{ coulombs/kg}$$

Notice that in this ratio of unlike quantities, the units do not cancel out. The interpretation of the ratio must also be slightly different from those in the examples considered previously. Here, we must state that the numerical size of the charge is 1.758×10^{11} times larger than the numerical size of the mass, in the units used. If the units are changed, the ratio will be changed also. In this example, when another independent experiment gave a measurement of the charge on the electron as $e = 1.60 \times 10^{-19}$ coulombs, it was then possible to use this result along with the preceding ratio to actually "weigh" the electron—to find its mass. In the equation $e/m = R$, solving for m, one has $m = e/R$. Thus, if the charge is divided by the ratio R, the result is the electron's mass. Doing this arithmetic, one has

$$m = 9.108 \times 10^{-31} \text{ kg}$$

which is an extremely small mass.

PROBLEMS

1. The distance from the planet Pluto to the sun is 5.908×10^{12} m. If the distance from the earth to the sun is 1.495×10^{11} m, what is the ratio of Pluto-sun distance to earth-sun distance?

2. A room has a length which is three times larger than its width. What is the ratio of length to width for the room? What is the ratio of width to length for that room?

3. A right triangle has legs of length $\sqrt{3}$ and 1. It has a hypotenuse of length 2. What is the ratio of the smaller leg to the hypotenuse? What is the ratio of the other leg to the hypotenuse?

4. For the following values of X and Y, find the ratio $R = X/Y$.
   a. $X = 4, Y = 2$ b. $X = 2, Y = 4$ c. $X = 5.43, Y = 7.98$
   d. $X = 1.18 \times 10^{-6}, Y = 5.59 \times 10^{-12}$ e. $X = 3, Y = 1/5$
   f. $X = 1/7, Y = 8$

5. Tables of atomic masses show the ratio of the mass of an atom of some element to the mass of hydrogen. Using a mass spectrometer, it is rather easy to compare masses of other elements with that of carbon. The spectrometer gives a ratio of some unknown mass to that of carbon. For example, the ratio of the atomic mass of silver to that of carbon is 8.99. If some independent measurement determines that the carbon atom has a mass of 1.99×10^{-26} kg, what must be the mass of the silver atom?

6. Change the following statement to symbolic form. The mass of the planet Jupiter is 318 times larger than that of the earth (Use M_e for mass of the earth, M_j for mass of Jupiter, and R for ratio).

6-2 The Meaning of Proportionality

If the distance "around" a circle, its circumference, is divided by its diameter, a ratio is formed which can be designated by C/D, where C stands for the circumference and D stands for the diameter. Circles of many different sizes can be drawn using compasses. For each of these circles, a piece of string can be placed along the circumference, cut to fit around the circle once, then straightened and placed next to a measuring scale. Each diameter can be measured by using the scale to measure the distance across the circle, passing through the center. For every circle that is constructed, when these measurements are carefully made, the ratio C/D turns out to be the same—a number of about 3.14. As discussed previously, the mathematician has found that this ratio is an irrational number called *pi*. Its approximate value is 3.1415926. For any circle, then, it can be said that the circumference will always be pi times larger than the diameter.

In electricity, when a certain voltage is pushing electric current through a resistance element, the ratio of voltage V to current I can be found. One can use a fixed resistance element for which changes in voltage give interesting changes in current. When the voltage is changed, the current always changes in such a way that the ratio re-

[§ 6-2]   The Meaning of Proportionality   83

mains constant. V/I is equal to a constant called R, the fixed resistance of the circuit.

In both of the examples discussed above, the ratios of two quantities always give the same number, a constant. When this result occurs, that is when all ratios are equal, the two quantities forming the ratio are said to be *proportional*. Using this word, the circumference C of a circle is proportional to the diameter D. In an electric circuit with a fixed resistance, the voltage V is proportional to the current I.

From the above discussion, one can specify a definition of the word proportional. Two quantities Y and X are said to be *proportional* if, for every value of one of the variables selected, the other variable has a value such that the ratio Y/X is the same, a constant. Symbolically, this proportionality would be written $Y/X = K$. When two quantities are proportional, the scientist or engineer often represents this fact by writing the two symbols representing the quantities separated by a wiggly line, like this: $Y \sim X$. This strange looking statement is read "Y is proportional to X", and it carries precisely the meaning that has been given to the word *proportional* in this chapter.

Using the above definition of proportional, one can deduce certain interesting properties of proportions. Consider the example of the circle. The circumference C is proportional to the diameter D. Symbolically, we would say $C \sim D$. The definition of proportional tells us that for any values of C and D their ratio is always a constant. If this is true, then when D is made twice as large, C must also be twice as large. If D is increased by a factor of three, C must also be increased by a factor of three. Using numerical examples to illustrate these facts, when D is 2 in., C is 6.28 in. When D is made twice as large to $2 \times 2 = 4$ in., C becomes twice as large to $2 \times 6.28 = 12.56$ in. When D is increased by a factor of three to $3 \times 2 = 6$ in., C increases by the same factor to $3 \times 6.28 = 18.84$ in. These illustrations point out a most important property of proportions: when one of the quantities is increased (or decreased) by some factor, the other quantity is increased (or decreased) by the same factor. In the example of the electrical circuit where $V \sim I$ (where the voltage V is proportional to the current I), if the current is increased by a factor of 20, then the voltage must also increase by a factor of 20. This is the only way in which the ratio of V to I can remain constant.

In the simple proportionalities discussed so far, the quantities taken as ratios have been to the first power, Y/X, C/D, and V/I. When proportions are of this type, it is said that one variable varies directly as the other. One would say that C varies directly with D, or V varies directly with I. Sometimes these particular first power proportions are called *direct proportions*. If the relationship between variables U and V is a direct proportion, the meaning of that statement is simply that U/V ratios are constant. The variables U and V are proportional.

There are proportions in science and technology for which the quantities are not first powers; they may have exponents. Such proportions are quite difficult to identify from tables of numbers, but there are certain techniques using special graphical methods that reveal such proportions quickly. The general definition of a proportion must take these into account; therefore, as a general definition, two functions, $f(x)$ and $f(y)$, are said to be proportional if for every value of x there exists a value of y such that

$$f(x)/f(y) = K$$

where K is a constant. Unless one has considerable mathematical sophistication, that definition will not mean very much, but it will be used in a restricted sense in some of the examples which are presented later in this chapter.

There is one important case of proportionality involving quantities which are *not* both to the first power. There are many variables in science and engineering whose product is a constant. Stated symbolically, $YX = K$ represents this relationship. To show the relationship as a ratio, we must write it in a slightly different form, $Y/(1/X) = K$. As a quick check on this representation, we should recall that to divide Y by $(1/X)$ one must "invert the divisor and multiply", thus obtaining $(Y)(X/1) = K$. From the previous definition of negative exponents, $1/X$ is the same as X^{-1}; therefore, the ratio form of this proportion can be stated as $Y/X^{-1} = K$. It can be said that Y is proportional to "one over X". This type of proportion is given a special name; it is called an *inverse proportion*. The quantity Y is said to vary inversely with the quantity X.

An example of an inverse proportion can be found in an electric circuit where the voltage is held fixed. An experiment with such a simple circuit shows that the product of the current I and the resistance R remains fixed as the resistance is changed. If the current is 5 amperes when the resistance is 20 ohms, the current must be 25 amperes when the resistance is 4 ohms. The product must always remain the same; in this case it is 100. Stated as a proportion, $I \sim R^{-1}$, or $I \sim 1/R$, or $I/(R^{-1}) = V$, where V is a constant. Regardless of how this inverse proportion is written, the important thing is its meaning. When one variable is increased by some factor, the other variable must be decreased by the same factor.

[§ 6-3]   The Proportionality Constant   85

PROBLEMS

1. If $R \sim T$, and $R = 5$ when $T = 8$, what will be the value of R when T is made twice as large?

2. An electronics technician notices that in a certain electric circuit the voltage across the circuit divided by the electrical resistance is always the same number. If the voltage is 100 when the resistance is 8, what would the voltage be when the resistance was 24?

3. An experiment involving gases at constant pressure shows that as the volume of the gas changes, the absolute temperature changes in such a way that the ratio V/T remains constant. If the temperature is made 1/3 as large, how much, and in what way, will the volume change? How is V related to T?

4. If $S \sim P^{-1}$, and $S = 4$ when $P = 9$, what will be the value of S when P is made twice as large?

5. An electronics technician measures the electrical current and resistance in a circuit where the voltage is held fixed. He notices that the product of current I and resistance R is always the same. If the current is 12 when the resistance is 5, what would be the current when the resistance was 15?

6. An experiment involving gases at constant temperature shows that as the volume V of the gas changes, the pressure P changes in such a way that the ratio P/V^{-1} remains constant. If the volume V is made 1/3 as large, how much, and in what way, will the pressure change? How is P related to V?

6-3 The Proportionality Constant

In the various examples of proportions discussed in the preceding section, the ratio of two quantities remained constant, or unchanging. This property was the essential feature of what was called a *proportion*. Proportions can always be expressed as a ratio equal to a constant number, which is called a *constant*. The mathematician gives this kind of constant a special name. Because it is an essential part of a proportion, the constant is called a *proportionality constant*.

In the ratio of the circumference of a circle to its diameter, the proportionality constant is equal to pi. In the example from electric circuits, where voltage V is proportional to current I, the proportionality constant is the electrical resistance R. In these and other such examples, the proportionality constant is the constant ratio that occurs if two quantities are proportional.

Using the proportionality constant, it is possible to write proportions in a simple and useful way. In order to test for a proportion, one must form ratios from all available data. If Y represents one quantity and X the other quantity, Y_1, read "Y sub one", represents one particular measurement of Y. The quantity Y_2 would represent another such Y value, but a different one from Y_1. This use of subscripts can be extended to any number of measurements of Y or X. Using these subscripts, a proportion can be defined as

$$Y_1/X_1 = Y_2/X_2 = Y_3/X_3 = Y_4/X_4 = \ldots Y_i/X_i = \ldots$$

That chain of equations simply says that all ratios are equal to each other, since they are all equal to the same number, the proportionality constant. The advantage of this definition of a proportion is that it describes how to test a set of numbers for proportionality. If every Y value divided by the corresponding X value is equal to every other such ratio, then the quantity Y is proportional to the quantity X. As an example, examine the following set of numbers.

R	T
9.6	3
6.4	2
16.0	5
22.4	7

These numbers can be tested for proportionality using the definition of proportion just given. First, the ratio $9.6/3 = 3.2$ is found. Then, if the other three ratios are equal to this value 3.2, and if all other measurements of R and T would give the same ratio, R would be proportional to T. Dividing the other three values of R by corresponding values of T does, in each case, give a result of 3.2. The quantity R is therefore proportional to T, and we can write $R \sim T$. The proportionality constant is 3.2.

Once an analysis of a set of data shows a proportionality and gives the value of the constant, the data can be summarized with an equation. Because the ratio of every pair of values is equal to the same constant, this fact can be stated by the equation

$$Y/X = K$$

or, multiplying both sides of this equation by X,

$$Y = KX$$

where K is the proportionality constant. In the preceding example, the analysis showed a direct proportion with $K = 3.2$; therefore, we would have the equation

$$R = (3.2)T$$

This equation summarizes the tabular data, but it also implies that the same proportionality exists for every set of values of R and T between and beyond the ones listed. Usually, this interpretation is valid for the in-between values, but the values beyond the ones observed may very well be related in quite a different way.

The preceding example was a direct proportion. The quantity R varied directly with T. As an example of an inverse proportion, consider the following set of measurements.

S	R
3.40	40
19.43	7
3.89	35
11.33	12
7.56	18

When a person begins to analyze such data, he does not have any idea what, if any, relationship exists between the two quantities S and R. To decide whether to test for a direct or an inverse proportion, he should examine the data to see how S changes with R. If S gets larger as R gets larger, the relationship could be direct. But if S gets smaller as R gets larger, then the relationship must in some way be inverse. With these data, as R becomes larger, S becomes smaller; therefore, the relationship is inverse. To determine what specific relationship exists, more must be done with the numbers than just looking at them. To test for an inverse proportion, we must divide S by the quantity $1/R$. The result is then the proportionality constant. But, as was shown previously, this ratio S/R^{-1} is the same as the product SR; thus we need only multiply the corresponding values of S and R in the table to test for an inverse proportion. If all products are the same, that number is the proportionality constant where S is inversely proportional to R.

The product of the first pair of numbers in the table is $(3.40)(40) = 136$. The second product is $(19.43)(7) = 136$, the same result as for the first pair. The product of each of the next three pairs of numbers is also 136. This relationship can be written as $S {\sim} R^{-1}$, or

$S \sim 1/R$, or, using the proportionality constant, $S = (136)(1/R)$. This equation could be more simply written as $S = 136/R$.

These two examples show how data can be analyzed to determine if a direct or inverse proportion exists, and how to calculate the proportionality constant. There are proportions, both direct and inverse, which are not nearly this simple. In those cases, it can be a very difficult task to manipulate the numbers until the particular relationship is discovered. It is also possible that no relationship exists which can be put into the form of a proportion.

PROBLEMS

1. When an airplane moves through the air at a speed V, it is observed that at low speeds the air resists the motion of the plane in a certain way. An experiment shows that the air resistance R is related to the airplane's speed V as shown in the following table.

R (in pounds)	V (in mph)
300	20
450	30
600	40
750	50
900	60

    What is the relationship between R and V? What is the proportionality constant? Write an equation for this relationship.

2. When weights are used to stretch a spring, the spring stretches according to the following table. What is the relationship between the weight W and the distance S that the spring is extended? What is the proportionality constant?

W (pounds)	S (inches)
0.3	2.00
0.4	2.67
0.5	3.33
0.6	4.00
0.7	4.67

3. The following sets of numbers may be related by a direct proportion, by an inverse proportion, or may have no simple relationship. Determine, in each case, whether the relationship is *direct proportion, inverse proportion,* or *no simple relationship.* When there is a simple relationship, com-

pute the proportionality constant and write the relationship as an equation.

a.

Y	X
4	1
8	2
12	3
16	4

b.

R	T
18	6
9	3
15	5
96	32

c.

U	V
1	1
7	1
1	8
2	16

d.

R	L
1.00	1
0.50	2
0.33	3
0.25	4

e.

X	Y
25	4
5	20
20	5
4	25

f.

T	L
0.0986	7
0.1151	6
0.1380	5
0.1725	4

g.

M	L
4	8
3	1
2	7
1	2

6-4 Ratio and Proportion in Similar Triangles

In a large number of scientific and technical problems and explanations, figures or diagrams which involve triangles are used. Sometimes relationships can be established between quantities represented by sides of similar triangles. In other cases, ratios of sides of right triangles are important. The concept of ratio and proportion must therefore be applied to triangles.

You will recall from your high school plane geometry course that when the angles of one triangle are equal to the angles of another triangle, these two triangles are said to be *similar*. In a geometry course, you would proceed to prove that, for similar triangles, corresponding sides are proportional. We shall not attempt to repeat that geometrical proof here. If you want to be convinced of its validity, you should refer to a textbook on plane geometry. But we should examine what it means to say that corresponding sides are proportional.

The triangles of Figure 6-1 are different in size, but corresponding angles of the triangles are equal. Sides of the triangles are marked off with scale divisions. *Corresponding sides* are those sides located between the same angles in each triangle. For example, side $\overline{AB}$ of the first triangle corresponds to side $\overline{DE}$ of the second triangle, side $\overline{BC}$ corresponds to side $\overline{EF}$, and side $\overline{AC}$ corresponds to side $\overline{DF}$.

A theorem from plane geometry states that these corresponding sides are proportional. From our definition of proportion, we know that corresponding sides must have the same ratio, and that the proportionality constant tells us how much bigger one triangle is than the other in terms of its linear dimensions.

From the scale marks in the figure, we can determine the proportionality constant. Reading the scale marks for the two triangles, and forming a table of data, we have the following set of numbers.

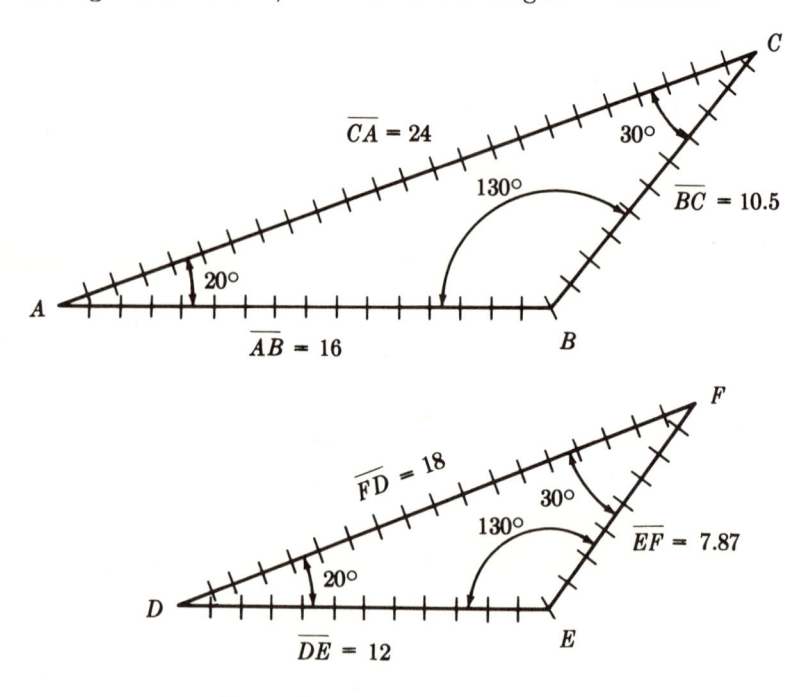

Figure 6-1. *Two similar triangles.*

Corresponding sides

Triangle ABC	Triangle DEF
$\overline{AB}$ = 16	$\overline{DE}$ = 12
$\overline{BC}$ = 10.5	$\overline{EF}$ = 7.87
$\overline{CA}$ = 24	$\overline{FD}$ = 18

These numbers can be analyzed just as was done in the preceding section. When ratios are formed, the results are 16/12 = 1.33, 10.5/7.87 = 1.33, and 24/18 = 1.33. The proportionality constant must then be 1.33, and corresponding sides are proportional. The linear dimensions of the larger triangle are 1.33 times larger than the corresponding dimensions of the smaller triangle.

The use of similar triangles in science and technology is usually in a derivation of some sort. A correspondence between the sides of two triangles and certain physical quantities is established. Then it is shown that the angles are all equal. From a knowledge of similarity, one can then set up proportions among the corresponding sides. This process usually produces a relationship that could not have been found in another way.

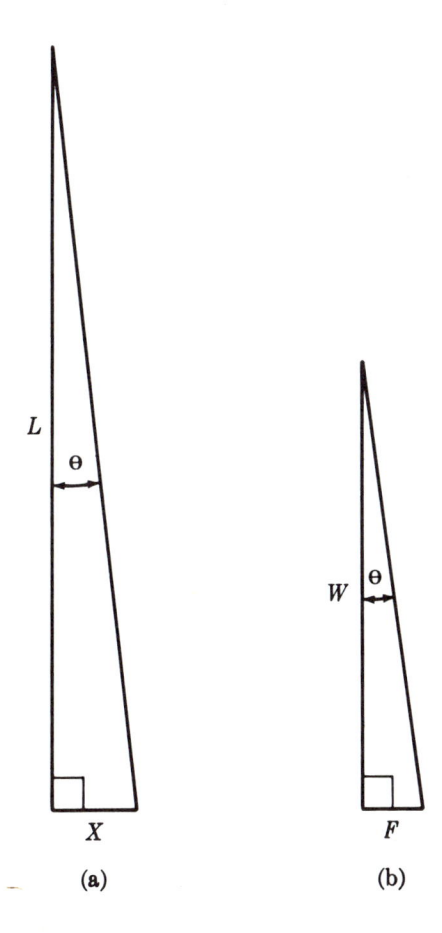

Figure 6-2. *A pendulum bob displaced a small distance from equilibrium.*

As an example, in the derivation of a formula for the force tending to restore a pendulum bob to its equilibrium position, the triangles of

Figure 6-2 would be used. The angle θ is assumed to be very small. The triangle in Figure 6-2(a) represents a pendulum with L its length and x its small horizontal displacement. The triangle in Figure 6-2(b) represents the forces on the pendulum bob, where W is its weight, and F is the horizontal restoring force.

The person studying this example knows from his previous studies that all angles are equal. Assuming that they are equal, he can use the property of similarity to set up proportions among the quantities described. If corresponding sides are proportional, their ratios must be equal to the same constant; therefore, those ratios must be equal to each other. Using that fact, $W/L = F/X$, and, solving this equation for F, one has $F = WX/L$, a formula for the force tending to restore the pendulum bob to its equilibrium position (assuming θ is very small).

The most common use of ratio and proportion involving triangles relates to the right triangle. Of course, a *right triangle* is a triangle containing one 90° (right) angle. To talk about the sides and angles of a right triangle, we use the words *opposite* and *adjacent*. For our purposes, opposite shall mean *across from*, and adjacent shall mean *next to*. Because we shall be talking about right triangles, the two shorter sides will be called *legs*, and the longer side will be called the *hypotenuse*. Figure 6-3 shows a right triangle with legs and hypotenuse indicated.

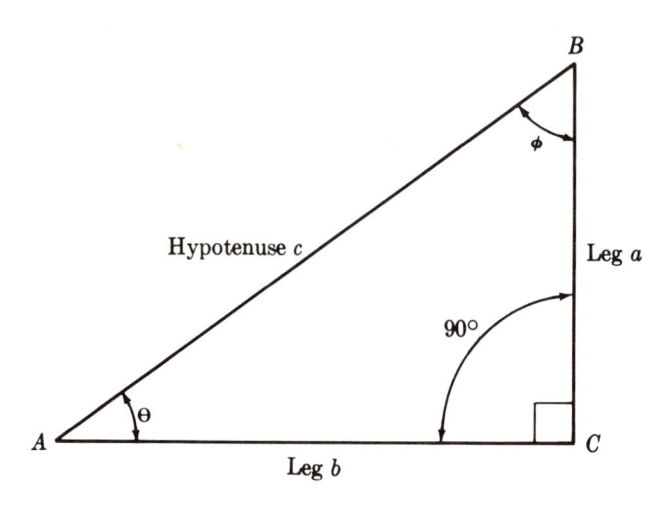

Figure 6-3. *The right triangle contains one 90° angle.*

Although not concerned with proportions, the relationship between the legs and hypotenuse of a right triangle is very important, and is used extensively in scientific and technical subjects. You probably

[§ 6-4] *Ratio and Proportion in Similar Triangles* 93

recall that relationship, the well-known *Pythagorean Theorem*: The sum of the squares of the legs of a right triangle is equal to the square of the hypotenuse. Stated symbolically, that theorem would be

$$a^2 + b^2 = c^2$$

Look at the triangle of Figure 6-3 again. Let us use the words *opposite* and *adjacent* to better understand their meaning. Leg b is opposite angle ϕ, but leg b is adjacent to angle θ. Notice that leg a is opposite angle θ, but adjacent to angle ϕ.

Figure 6-4 shows two right triangles, the smaller one formed by erecting a perpendicular from the base of the larger. This perpendicular can be placed anywhere along the larger base; its location is arbitrary for the discussion that follows.

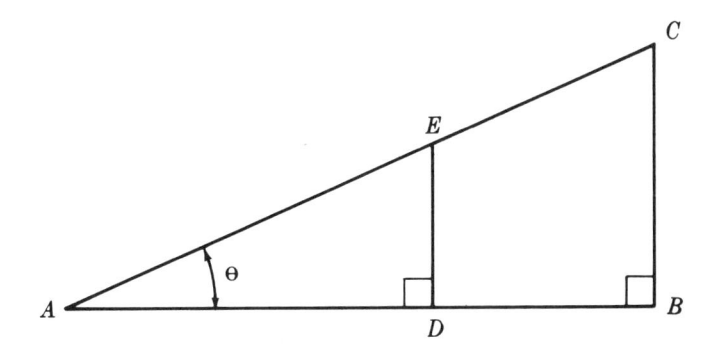

Figure 6-4. *Two right triangles, each having a common angle.*

The two triangles, ABC and ADE, share the angle θ. Each also has one right angle; therefore, the other corresponding angles are equal. All angles of triangle ABC are equal to corresponding angles of triangle ADE. For that reason, the two triangles are similar, and corresponding sides are proportional.

To say that corresponding sides of the two triangles are proportional requires that the ratio $BC/AC = DE/AE$. You can measure those distances in Figure 6-4 with a scale and verify that the two ratios are equal. Notice that whether we are talking about triangle ADE or triangle ABC, if we use the words "leg opposite angle θ divided by the hypotenuse" we get ratios that have the same value. In fact, if we just had one right triangle of *any* size, so long as the angle θ remained the same, the ratio "leg opposite angle θ divided by hypotenuse" would always be the same. So would any of the other side ratios. Because the ratio of each leg of a right triangle to the other leg and to the hypotenuse is the same number for a given angle, regardless of the size of the triangle, these calculated ratios might be useful if we knew them for all acute angles. Such ratios have been determined and put into tables. They are often found in the appendix of a science or engineering textbook. They are called *trigonometric ratios*.

Looking again at Figure 6-3, you will notice that a right triangle has six possible side ratios. If we use the angle θ as a reference, the ratios would be as follows:

> Leg opposite θ divided by the hypotenuse
> Leg adjacent to θ divided by the hypotenuse
> Leg opposite θ divided by leg adjacent to θ
> The hypotenuse divided by the leg opposite θ
> The hypotenuse divided by the leg adjacent to θ
> Leg adjacent to θ divided by leg opposite θ

Because these ratios are constant for a given angle θ, regardless of the size of the right triangle, they are each given a name, just as we name a person Bill or Jim. The names of these constant ratios are as follows:

> sine θ = leg opposite θ/hypotenuse,   abbreviated sin θ
> cosine θ = leg adjacent to θ/hypotenuse,   abbreviated cos θ
> tangent θ = leg opposite θ/leg adjacent to θ,   abbreviated tan θ
> cosecant θ = hypotenuse/leg opposite θ,   abbreviated csc θ
> secant θ = hypotenuse/leg adjacent to θ,   abbreviated sec θ
> cotangent θ = leg adjacent to θ/leg opposite θ,   abbreviated cot θ

The most commonly used trigonometric ratios are $\sin\theta$, $\cos\theta$, and $\tan\theta$, especially since one can find the other three simply by taking reciprocals. These names are read "sine of theta, cosine of theta, and tangent of theta", respectively. One should commit the definitions of these three trigonometric ratios to memory. They are used extensively in science and technology. They should be memorized in terms of the words "opposite", "adjacent", and "angle θ", not in terms of letters like a, b, or c. There are many relationships among the various trigonometric ratios, but you should learn those identities in a course in trigonometry, or, if necessary, consult a text on trigonometry.

PROBLEMS

1.   Triangle ABC is similar to triangle DEF (the triangles have equal angles). Side $\overline{AB}$ corresponds with side $\overline{DE}$, side $\overline{BC}$ with $\overline{EF}$, and side $\overline{CA}$ with $\overline{FD}$. You are given that $\overline{AB} = 4$, $\overline{BC} = 9$, and $\overline{CA} = 7$. If side $\overline{DE}$ of the triangle DEF is 16, what is the length of $\overline{EF}$ and $\overline{FD}$?

[§ 6-4] *Ratio and Proportion in Similar Triangles* 95

2. By several different kinds of measurements, we know that the distance to the moon is 3.84×10^8 m (239,000 mi). If one fastens a quarter to a window pane with rubber cement, it can be used to measure the moon's diameter. With the full moon visible outside this window on the horizon, close one eye and slowly walk forward until the quarter just blocks out the view of the moon's disc through your other eye. Then measure the distance from your eye to the quarter. When this procedure is followed, you will get an eye to quarter distance of 2.645 m. Using the fact that the quarter has a diameter of 0.024 m, construct a diagram of two similar triangles, and use a proportion to find the diameter of the moon.

3. Find the indicated trigonometric ratio in each of the following triangles. If a leg or hypotenuse of a triangle is unknown, but needed, use the Pythagorean Theorem to determine what it should be.

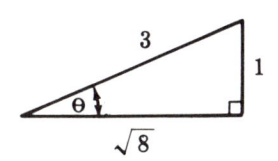

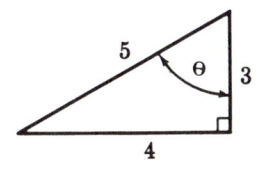

a. Sin θ

b. Cos θ

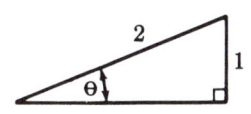

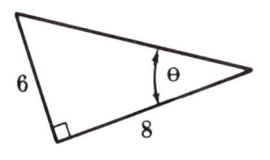

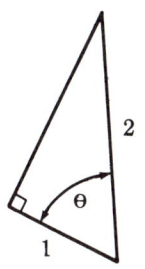

c. Tan θ

d. Sin θ

e. Cos θ

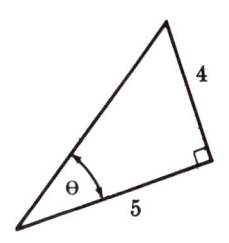

f. Tan θ

4. A cannon fires a shell at a speed of 837 ft/sec at an angle of 60° measured from the horizontal. Use the trigonometric ratio cos 60° to find the initial speed of the shell in the horizontal direction. A table of trigonometric ratios would give cos 60° = 0.5.

6-5 Examples of Ratio and Proportion from Science and Technology

The uses of ratios in science and technology are of two types. In the first, quantities of a different kind are taken as ratios. In the second, quantities of the same kind are formed as ratios. The ratio of charge to mass for the electron has already been discussed. It was seen how this ratio could be determined in a given experiment; then a later experiment measured the charge, so that the combination of these two experiments "weighed" the electron.

A ratio of mass to volume gives a quantity called *density*. Stated symbolically, the ratio $m/V = \rho$ measures the "heaviness" of some substance. For example, although Jupiter is much larger than the earth and "weighs" more, having a mass of 1.9×10^{27} kg, compared with 5.98×10^{24} kg for the earth, the density of Jupiter is only 1.33×10^3 kg/m³, compared with a value of 5.52×10^3 kg/m³ for the earth. This density difference means that the earth is made of "heavier" stuff than Jupiter is.

Proportions are much more commonly used in science and technology than simple ratios. Proportions have the advantage of providing for easy solution to certain problems when only two quantities change, all else remaining constant. For example, Hooke's Law states that the force needed to extend or compress a spring is proportional to the extension or compression. Symbolically, this proportion is written $F \sim x$. Without even knowing the proportionality constant, you can form a useful equation from this proportion. Because the proportion exists only if all ratios of F/x are constant, and therefore equal to each other, you can consider two different conditions of the same spring and use subscripts to indicate the values of F and x. The proportion can be written as

$$F_1/x_1 = F_2/x_2$$

Then, when the force and extension are known in one case, and either of the variables is known in the second case, you can compute the unknown quantity. If the extension is 4 in. when the force is 8 p, then when the extension is 12 in. the force must be given from the equation

$$8 \text{ p}/4 \text{ in.} = F_2/12 \text{ in.}$$

Solving this equation for F_2, you get a force of 24 p.

From this example, we can describe a procedure for converting a proportion, $Y \sim X$, into a useful equation. First it must be assumed that all variables remain the same except Y and X. Then the proportion definition requires that Y/X be the same for all corresponding pairs Y and X. Thus, if we call one pair Y_1 and X_1, and another pair Y_2 and X_2, we must have the equation

$$Y_1/X_1 = Y_2/X_2$$

This equation can then be used to solve problems.

A similar procedure can be used for inverse proportions, where Y/X^{-1} is the constant ratio. As shown previously, this proportion requires that the product of Y and X be constant. If the pair Y_1, X_1 represents the first situation, Y_2, X_2 can represent the changed set of quantities. But because the products must be constant, and therefore equal to each other, we have the equation.

$$X_1 Y_1 = X_2 Y_2$$

An example of this type of relationship exists with gases at constant temperature. The principle is Boyle's Law, where $P \sim 1/V$, an inverse proportion. To make this law useful, we write it as an equation:

$$P_1 V_1 = P_2 V_2$$

Of course, its validity depends upon all other variables remaining constant. If the temperature were to change, this "law" would be invalid.

The great advantage in the use of proportions is being able to convert a complicated equation into the form of a proportion. This process is relatively simple, and often makes possible rapid solution of otherwise difficult problems. Consider the following example. $F = GmM/R^2$ is the law of universal gravitation. In this law, G is a universal constant, a constant of proportionality. The quantities m and M are masses of two bodies, and R is the distance between their centers. As an equation, this statement says to multiply G, m, and M together, dividing the result by the square of R; but it also consists of several possible proportions. If m and R are constant, it says that $F \sim M$. If M and R are constant, it says that $F \sim m$. If both m and M are constant, it says that $F \sim 1/R^2$. What is shown by these examples is that any quantity in the numerator is related as a direct proportion, and any quantity in the denominator is related as an inverse proportion. The type of proportion is also dependent upon the power of the quantity involved, like R^2. These facts make it possible to construct proportions from equations to solve problems involving only two variables.

For example, assume that in the law of universal gravitation the mass m is that of a satellite, M is the mass of the earth, and we want the force on the satellite at different distances from the center of the earth. The problem would be easier to solve as a proportion, where $F \sim 1/R^2$, than as an equation where substitutions must be made for all the quantities in just the right units. Figure 6-5 shows the two situations. By a procedure similar to that used for an inverse proportion, this inverse square proportion can be made into an equation by setting ratios F/R^{-2} equal to one another. But F/R^{-2} is, by the definition of the negative exponent, just FR^2. Thus we can write the equality

$$F_1 R_1^2 = F_2 R_2^2$$

Then, in using this proportion, we must be careful to square the values of R. As an example, let us compute the force on a satellite at a distance of two earth's radii from the earth's surface, if the force at the earth's surface (R = one radius) is 400 p. For the first situation, $F_1 = 400$ p and $R_1 = R_E$, where R_E is the radius of the earth. In situation number two, the satellite is at $3R_E$ and the force is F_2. Writing the proportion, we have

$$(400)(R_E)^2 = F_2(3R_E)^2$$

Squaring the indicated quantities, and dividing both sides of the equation by R_E^2, we have

$$400 = F_2(9)$$

or

$$F_2 = 400/9 = 44.4 \text{ p}$$

This inverse square law is extremely important because it occurs for several different kinds of quantities in science and technology. Light and sound, although different, are both propagated from a point source according to the inverse square law. Electrical forces from point charges follow the inverse square law. These different laws become identical in form when expressed as proportions; thus the importance of the proportion form of the inverse square law cannot be overly stressed.

Proportions may be of many different kinds, but those most common in engineering or science are the direct proportion, where

$$Y_1/X_1 = Y_2/X_2$$

the inverse proportion, where

$$Y_1 X_1 = Y_2 X_2$$

and the inverse square proportion, where

$$Y_1 X_1^2 = Y_2 X_2^2$$

[§ 6-5]   Examples of Ratio and Proportion   99

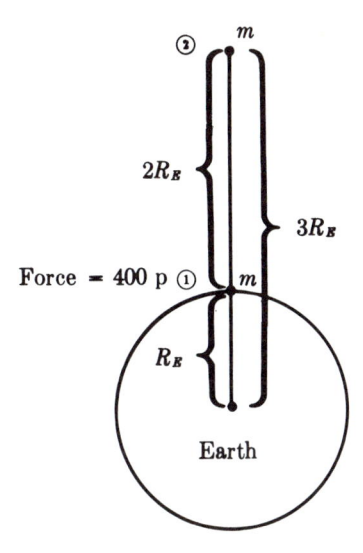

```
            ② ┌ m
              │ │
       2R_E  ⎰│ │
              │ │ ⎱
              │ │  ⎰ 3R_E
Force = 400 p ① ⎱│m│ ⎱
              ┌─┴─┴─┐
             ╱  │ │  ╲
            │ R_E │   │
             ╲  └─┘  ╱
              ╲Earth╱
               ╲___╱
```

Figure 6-5.   *A satellite at the earth's surface and at a distance of $2R_E$ from the earth's surface.*

You should have the ability to convert a proportion statement into one of the above forms from your understanding of the definition of proportion. You should also be able to look at an equation and state what proportions would exist under certain conditions of constant quantities.

PROBLEMS

1.   The mass of Saturn is 5.69×10^{26} kg, and its radius is 5.75×10^7 m. Using the formula for the volume of a sphere, $V = (4/3) \pi R^3$, compute the volume of Saturn; then find the ratio m/V, the density of Saturn.

2.   Ohm's Law in electrical circuits is stated by the equation $V = IR$. If a certain circuit has a constant resistance R, then $V \sim I$. Change this proportion into an equality involving subscripts like 1 and 2. If the voltage is 6 when the current is 9, use the equation just formed to find the current when the voltage is 18.

3.   If $F = ma$ is an equation, prove that $F \sim a$ when m is constant using the definition of a proportion. Use that same definition to prove that when m is constant, $F_1/a_1 = F_2/a_2$.

4. In physics there is a law relating the volume V of a gas and the absolute temperature (centigrade temperature plus 273) when the pressure of the gas is constant. This relationship, called Charles' Law, can be stated by the proportion $V \sim T$. Change this statement of the proportion to an equation using subscripts; then if the volume is 9 when the temperature is 300, find the volume when the temperature is 442.

5. The final energy E of a particle of charge q accelerated in a cyclotron is given by the formula $E = R^2q^2B^2/2m$, where R is the radius of the cyclotron, B is the strength of the magnetic field in the cyclotron, and m is the mass of the particle accelerated. Assuming other variables are constant, how is the energy related to each of the variables listed? As the magnetic field strength is increased by a factor of three, how much does the energy of the particle change, assuming all other factors remain constant?

6. The propagation of light from small sources follows the inverse square law, $I \sim 1/R^2$, where I is the illumination and R is the distance away from the light source. By spectral examination of the light from certain distant stars, it can be determined that those stars are in the same evolutionary stage as our own star, the sun. From this kind of evidence we can determine how intrinsically bright a distant star is (like determining whether a distant street light is 100 watts or 500 watts). By comparing the amount of light we receive from a star with that from our sun, we can use the inverse square law to "measure" the distance to that star. Make the above proportion statement into an equation involving subscripts; then compute the distance to a near star which has a measured brightness here of only 2.2×10^{-15} as bright as our sun. The distance to our sun is 1.5×10^{11} m.

7. In the following proportions, how much and in what way would the quantity on the left be changed if the quantity on the right is

a. increased by a factor of two?   b. decreased by a factor of two?
c. increased by a factor of three?   d. decreased by a factor of seven?

1. $F \sim X$
2. $P \sim V^{-1}$
3. $F \sim 1/R^2$
4. $E \sim V^2$
5. $U \sim S^{-3}$

6-6 General Ideas of Ratio and Proportion

The concepts of ratio and proportion are often extended and used in rather general ways to give curious results in certain areas of science and engineering. In the study of the biological cell, the ratio of surface area to volume becomes a determining factor relative to the cell size. In working with scale models, the engineer must be aware of how certain scale factors are related to the weight and strength of the structure being designed.

To understand the problems discussed above, certain kinds of proportions must be understood. For living systems, the heat produced by metabolic processes must pass through the surface of the organism; therefore, since surface areas are always calculated from formulas involving squares of linear dimensions, the proportion $H \sim L^2$ can be stated for any living organism. H is the heat passing through its surface and L is any linear dimension; it might be its length, width, or any imaginary line drawn through the organism. The constant of proportionality, of course, varies according to the shape and kind of organism. The food needed by a living organism is proportional to the amount of substance in the organism—to its volume. Since volume is proportional to the cube of linear dimensions, the food needed in the organism follows the proportion $F \sim L^3$. The food for an organism must pass through some surface, the stomach for a man, or the cell wall of an amoeba. But any such surface has an area which is proportional to the square of some linear dimension. As a consequence, the food available, A, for an organism follows the proportion $A \sim L^2$. From these two proportions it can be seen that as an organism gets larger and larger, its food gathering ability decreases in comparison with its food needs. It begins to starve.

The strength of any structure is proportional to the cross-sectional area of the supporting parts. Because areas are proportional to squares of linear dimensions, the proportion $S \sim L^2$ applies to the strength of a structure. But the weight of a structure is proportional to the volume, and volume is proportional to the cube of the linear dimensions. Thus the proportion $W \sim L^3$ applies to the weight of a structure. By taking the ratio of strength to weight of a structure, we get $S/W \sim 1/L$. As an object is "scaled up", made larger in all linear dimensions, its strength does not increase as fast as its weight. It would eventually collapse. But as an object is "scaled down", made smaller in all linear dimensions, its strength would get very large compared with its size.

As an example of the scaling laws discussed above, consider the possibility of a giant man on earth, say a man ten times larger in *every linear dimension*. He would be 60 ft tall, about 20 ft wide, and about 10 ft thick. He would be strong, all right; since $S \sim L^2$, he would be $(10)^2 = 100$ times stronger than a normal-size man. But his weight would be related by $W \sim L^3$; he would be $(10)^3 = 1000$ times heavier. He would be only 100 times stronger, but 1000 times heavier. How would he feel? It would be as though he were carrying 10 people his own size around on his back. His bones would probably be crushed by the excessive weight.

Of course such a giant creature would have other difficulties. His surface area and that of his stomach would be $(10)^2 = 100$ times larger, but the substance of his body that needs food would be $(10)^3 = 1000$

times larger. He would need 10 times more food than the surface of his stomach could permit to enter.

The examples discussed above merely serve to illustrate the general ways in which proportions can be used to find information about systems without knowing very much about the system. In the practical business of engineering, scaling factors must be considered in the use of scale models. A small scale model of an airplane is stronger for its size than the real thing. Measurements made on the scale model must be adjusted for the scale factor before being applied to the real airplane. There are many other instances where quantities may change in very strange ways as we scale to the very large or very small. In any case, the use of proportions may be essential when large numbers of variables are involved. Their use provides a powerful method of solving certain kinds of problems in science and technology.

PROBLEMS

1. How much less heat would be lost by a living organism made 100 times smaller in all linear dimensions? How much less heat would be produced by that organism? What kinds of problems would this creature have?

2. A movie shows a giant ant, 1000 times larger than a normal-size ant in all linear dimensions. How much stronger than a normal ant is this giant? How much heavier would this giant be than a normal-size ant? What kinds of problems would such a creature have?

3. An engineer working with a scale model of an airplane notices that the landing gear will just support the model without breaking. If this model is only 1/100 the size of a full sized aircraft in *all linear dimensions*, how much stronger is the larger landing gear than the smaller one? How much heavier is the larger aircraft than the model? What would have to be done to make the real airplane safe?

4. In making movies using small scale models of bridges, ships, etc., the timing of action sequences does not quite look right unless some special filming technique is used. The parts of a small model bridge exploding move too fast. A model ship pitches too fast, in comparison with a real one. To make such models look realistic, the period of time for some action T must be made proportional to the square root of the linear dimension L, $T \sim \sqrt{L}$. Using this proportion, how much faster must the filming of a model ship be than a real ship, if the model is 1/64 as large as the real ship in all linear dimensions? The film is then shown at the slower, normal rate, making the motion appear slowed by the same factor as the one computed.

7 Measurements of Physical Quantities

The reader has already learned how all physical laws, the principles upon which our modern technology is based, derive from only a small number of fundamental physical quantities: measures of space, time, mass, and electric charge. It would seem that measurement would be a process deserving little attention in studying science or engineering, since it appears such an easy thing to do. However, the processes of calibration, of establishing units, of counting units and subunits, and of estimating uncertainties are all part of making a measurement. For measurements which are built from two or more fundamental quantities, so-called *derived* measurements, the difficulties are even more severe.

7-1 Simple Measurements of Length, English and Metric Units

The process of length measurement was discussed in Chapter 1, where the unit of length was the "bleep". This unit was subdivided by tens, so that measurements could be expressed in the decimals of our number system. In terms of its subdivisions, the bleep is like the meter. As discussed in Chapter 1, the meter is subdivided into 10 parts called decimeters. The meter is also subdivided into 100 parts called centimeters, and into 1000 parts called millimeters. These millimeters are the smallest marks on the familiar meter stick.

The English unit is somewhat more difficult to use than the metric, primarily because the subdivisions are not usually based on tens. For example, the yard is divided into 3 ft, the foot into 12 in., and the inch is usually divided into eighths, sixteenths or thirty-seconds. To make a measurement of length in the English system requires that the sub-units be converted from fractional parts to decimals. For example, the width of a room is 12 ft, 4 and 5/32 in. To write this measurement as a decimal requires that 5/32 be changed to the decimal 0.1563. Then this 4.1563 in. must be changed to feet by finding 4.1563/12, which is 0.346 ft. Finally the measurement can be expressed as 12.346 ft. This example illustrates the difficulties of using the English system in comparison with a metric system where the sub-units automatically can be written as decimals without doing any arithmetic.

Figure 7-1.   *A metric scale in comparison with an English scale.*

You may ask, "Why don't we use the metric system instead of the English system, since it is so much easier to use?" The answer is not simple. Our entire industrial complex has been established using the English system. Tools and machines are calibrated in English units. Books and expensive references are written in terms of English units. Even our skilled workers in machine trades have learned to work only with the English system. It would be extremely difficult and expensive to make the conversion, even though our continued use of the English system is very expensive to us in a competitive world market. We are

[§ 7-1]   *Simple Measurements of Length, English and Metric Units*   **105**

stuck with the English system for a while. The only thing one can do is learn to use both systems of units, the English and the metric.

How is a length measurement made? What is actually done? Let us try to answer these questions in a general way, instead of talking about a particular measurement. The first requirement is something to be measured. There must be something about the object which we can identify as a starting point, and something else to be the end point of the measurement.

In measuring the width of a room, or of a book, or of some other such object, it seems an easy thing to identify the starting and ending points for our measurement. But if the measurement is to be very precise, perhaps where a microscope is used to locate those edges, they sometimes look quite fuzzy compared with the small subdivisions we might be using. Under such circumstances, the size of the smallest sub-unit possible, and therefore the precision of the measurement, is dependent upon the fineness of the edges of the objects we are trying to measure.

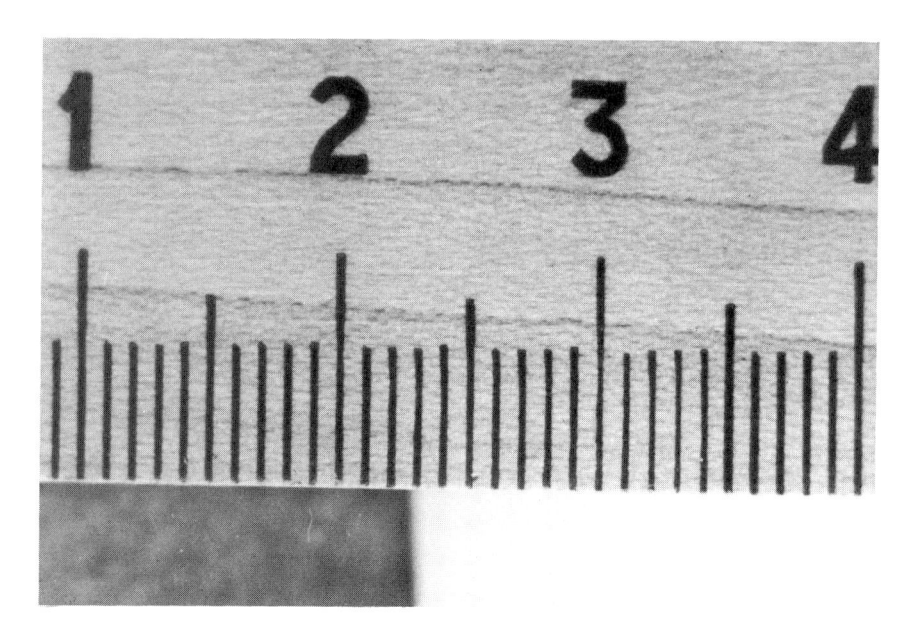

Figure 7-2.   *The fuzzy edges of the object to be measured are almost as large as the smallest scale division.*

The second requirement for making a measurement is to have a measuring instrument. If the measurement is to mean anything to anyone else, the instrument selected must have been calibrated with an established standard.

For the common foot ruler, yardstick, or meter stick, the manufacturer has compared his apparatus with a standard length at the Bureau of Standards. But if the instrument is constructed of wood, it is likely that expansions or contractions can change the length somewhat over a period of time. A stack of meter sticks in a college classroom might have lengths which vary one from another by as much as one or two millimeters. If a measurement is important, it may be necessary to go to a local Bureau of Standards and compare one's unit of length with a standard.

Once one has something to measure, and a calibrated instrument with which to perform the measurement, he is ready to begin counting units and sub-units. The process of measurement involves placing the measuring instrument with its initial end at the starting point of the object to be measured. Then the number of units of length that can be placed between the starting point and end point are counted. Usually, such a measurement gives a certain number of units, but with space left over, smaller than a full unit. For that space, sub-units are counted until the smallest subdivision observable has been considered. The process of making a length measurement is therefore one of counting units and sub-units.

Let us consider as an example the measurement of the length of a pendulum using a previously-calibrated meter stick. The starting point for our measurement is the point of support of the pendulum shown in Figure 7-3. Placing the meter stick with one end at that point, we notice that the length is greater than one unit. But the pendulum is shorter than two meters; therefore we must count sub-units. Counting sub-units of centimeters from the end of the meter stick to the center of the pendulum bob which forms our end-point, we count 22 sub-units, with a small space left which is smaller than a centimeter. We now must count from the end of the 22 centimeter sub-units to the end-point of our pendulum in smaller sub-units of millimeters, the smallest marks on the meter stick. We count 4 of these millimeters, with a very small space left, smaller than a millimeter. Because our meter stick has no other marks on it, and the millimeter is only about as wide as a pencil lead, we can only guess the next smaller sub-unit. In this measurement, the center of the pendulum bob looks like it is about half way between that 4th millimeter mark and the 5th millimeter mark. We must make a guess for the last digit. Let us select the 5th mark.

We have counted units and sub-units using the meter stick to measure the length of a pendulum. We counted 1 meter unit, 22 centimeter sub-units, and 5 millimeter sub-units. Because the metric system is

[§ 7-1] *Simple Measurements of Length, English and Metric Units* **107**

based on tens, we can write this measurement as 1.000 m plus 0.220 m plus 0.005 m, or 1.225 m.

Figure 7-3. *The meter stick is used to measure the length of a pendulum longer than one meter.*

If the same measurement had been made with a yardstick, we would get 1 yd, 1 ft, and 7/32 in. To use this measurement in formulas, it must be in decimal form. Thus we must convert 1 yd to 3 ft, add this result to 1 ft to get 4 ft. Then we must change 7/32 in. to a decimal fraction of a foot by dividing by 12. This division gives a result of 0.018, and our measurement is 4.018 ft, a somewhat more difficult measurement to make than the measurement in the metric system.

One must be able to convert from the English system to the metric system as needed. Today, the international standard unit of length is the meter. The English unit is defined in terms of the centimeter; one inch is exactly 2.54 cm. If a measurement must be converted from the English to the metric system, this one number must be remembered. It is the key to length conversions.

PROBLEMS

1. Express the following metric measurements in decimal form in terms of meters.
   a. 3 m, 48 cm, 7.3 mm   b. 0 m, 0 cm, 8 mm, 158 μ
   c. 9 km, 12 m, 5 cm, 8 mm   d. 0 m, 78 cm, 3.8 mm

2. Express the following English measurements in decimal form in terms of feet.
   a. 2 yd, 2 ft, 7 and 5/16 in.   b. 1 mi, 438 yd, 1 ft, 4 in.
   c. 0 yd, 8 ft, 11 and 22/32 in.   d. 0 yd, 5 ft, 3 and 17/64 in.

3. Convert the following English length measurements to metric units in meters.
   a. 1 yd   b. 1 ft   c. 1 mi   d. 3.84 ft   e. 6 ft, 8 and 3/16 in.

4. Convert the following metric measurements to English units in feet.
   a. 1 km   b. 1 m   c. 3.89 m   d. 73.12 cm

7-2 The Vernier Scale on Length Measuring Instruments

As shown previously, a measuring instrument with smallest subdivisions of millimeters, or sixteenths or thirty-seconds of an inch, requires a guess for the final digit of a measurement. However, for many such instruments a device can be attached to the scale which permits reading the space between smallest sub-divisions without actually having additional main scale marks on the instrument. This device is called a *vernier scale*.

As shown in Figure 7-4, the vernier scale has 10 spaces marked on it, and is placed next to the main scale of a measuring instrument. The vernier scale divisions are made so that ten of them just fit along 9 of the smallest main scale subdivisions. This ratio of scale divisions is just what is needed to provide a method of subdivision of the smallest main scale spaces.

Because 10 vernier spaces equal 9 smallest main scale spaces, if mark number N on the vernier scale just lines up with a mark on the main scale, the distance from that pair of aligned marks back to the

[§ 7-2]   The Vernier Scale on Length Measuring Instruments   109

zero vernier mark must be $(9/10)N$ smallest main scale spaces. The zero vernier mark would then be aligned between two main scale marks at $N/10$ from the left edge of that main scale mark. For example, if the fifth vernier mark is aligned with some mark on the main scale, the zero vernier mark must be halfway between two of the smallest main scale marks. If it were the third vernier mark, the zero vernier mark would be at $3/10$ of the way. From these examples, we can determine a method for reading between two main scale marks using the vernier scale.

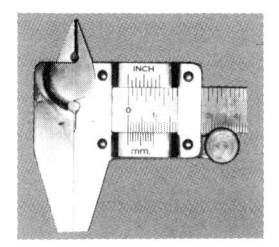

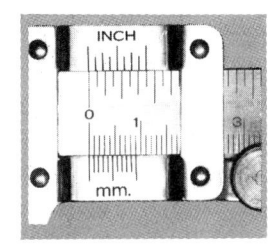

Figure 7-4.   *Ten vernier scale divisions are placed next to only nine main scale smallest divisions.*

An instrument with a vernier scale is constructed so that the left vernier scale mark is on zero when the measurement is zero. This fact, and what we have just learned about the location of that zero vernier scale mark, make it rather special. We shall call the zero vernier scale mark the *vernier index*. The vernier index will be used as an indicator to determine all digits of a measurement except the last, which will be read from the alignment of the vernier scale with the main scale. The last digit of a measurement, a fraction of the smallest main scale division, is determined by visually searching along the vernier scale until a vernier mark appears to line up with a main scale mark. The number of the aligned vernier mark is the correct fraction of the smallest main scale division. If the vernier has 10 spaces, that fraction is in tenths.

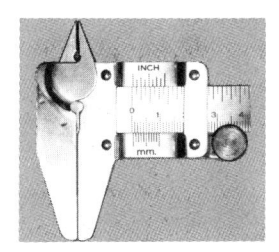

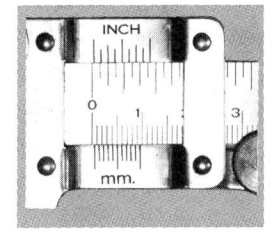

Figure 7-5.   (a) *A vernier scale with a reading of one-half the smallest scale division.*

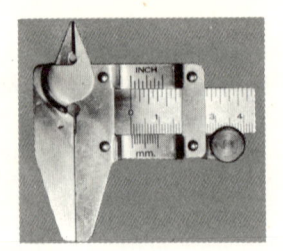

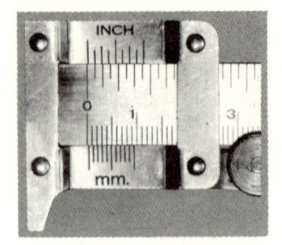

Figure 7-5. *(b) A vernier scale with a reading of three-tenths of the smallest scale division.*

Figure 7-6 shows an instrument with a vernier scale. The smallest main scale divisions are in millimeters, and the vernier scale has 10 spaces. What is the reading of the instrument? The vernier index falls between 1.2 cm and 1.3 cm. Without a vernier scale, we would guess the next digit, perhaps as 3 or 4. It would be difficult to tell. But with the vernier scale we can see that a vernier scale mark is aligned with a main scale mark at vernier mark number 3. Thus, the vernier index is 3/10 of the way between two millimeter marks, and our reading is 1.23 cm.

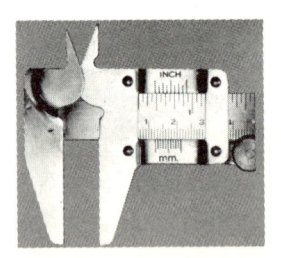

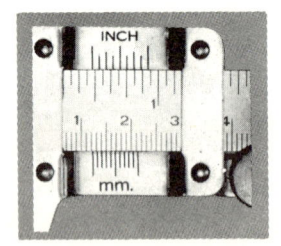

Figure 7-6. *Using the vernier scale, the measurement is 1.23 cm.*

Vernier scales are not always constructed with ten spaces. On an English scale, there might be eight vernier spaces corresponding with 7/16 in. This vernier scale would permit measurements to the nearest 1/128 in. with smallest main scale marks of 1/16 in. Some vernier scales have 20 spaces or more, permitting even finer measurements. There are other instruments than those for measuring length which use the vernier scale. For example, many spectrometers use a vernier scale for subdividing angular measure. But the principle is the same. If you understand how to use one kind of vernier scale, you can usually figure out how to use another that you may find on some other measuring instrument.

[§ 7-2]   *The Vernier Scale on Length Measuring Instruments*   **111**

PROBLEMS

1. In the following figures, read the measurement, finding the last digit by means of the vernier scale. Smallest units are millimeters.

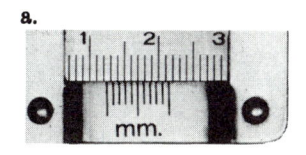

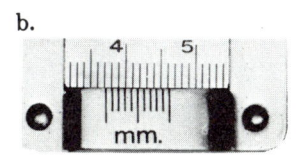

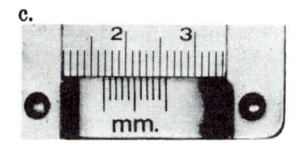

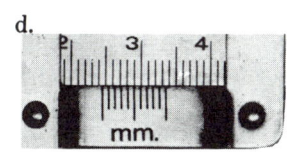

   a.

   b.

   c.

   d.

2. In the following figures, read the measurement, finding the last digit by means of the vernier scale. Smallest units are 1/16 in.

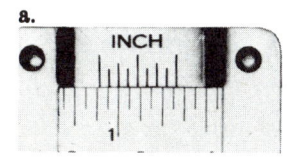

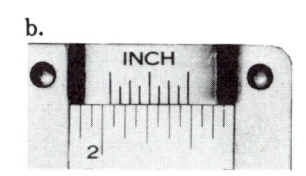

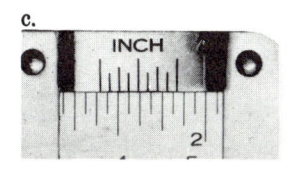

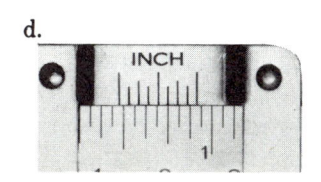

   a.

   b.

   c.

   d.

3. The following vernier scale has 20 spaces for 9 spaces along the main scale. What is the reading of the instrument?

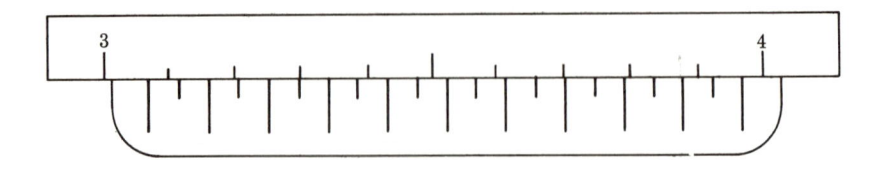

7-3 Simple Measurements of Mass, English and Metric Units

As has been discussed previously, the measurement of mass is a process of comparison. The instrument for making that comparison is called a balance. Comparing masses can be as crude and simple as "hefting" weights in your hand to judge whether they are the same, or as complicated as comparing masses in an electronic balance. The principle is still the same.

Before discussing mass measurement by the balance, we should make clear the differences between mass and weight, and between gravitational mass and inertial mass. The words mass and weight are used almost interchangeably by the chemist and by the engineer. The physicist is usually very careful to distinguish the two quantities.

Weight is the force with which gravity pulls on an object. This force is usually measured by spring scales. It is directed downward. In fact, we mean by *down* the direction in which gravity pulls something. The physicist has discovered that weight is proportional to mass, that is, $W \sim M$. The proportionality constant between weight W and mass M is a quantity called the *acceleration of gravity*, designated by the letter g. In the MKS system, weight W is measured in units called newtons, g is measured in m/sec², and mass M is measured in kg. In the familiar English system, weight W is in p, g is in ft/sec², and mass M is in units called *slugs*. The quantity g is the rate at which an object speeds up as it falls near the earth's surface. It has a value of about 9.8 m/sec/sec, or, 32 ft/sec/sec.

From the preceding discussion, we can see that although weight is proportional to mass, it is not the same thing. Technically, we should not confuse the two quantities. We are so accustomed to using the word "weigh" when we are to find the mass of something that it is difficult to divorce our thoughts about weight and mass. The best way to decide which is being done is on the basis of units. If the units are grams, kilograms, micrograms, slugs, etc., we are finding mass. But if the units are pounds, newtons, or some other such unit, we are really finding force, which we call weight. We might make a distinction on the basis of the instrument used. If a balance is used, we are comparing masses, but if spring scales are used, we are measuring weight. Unfortunately, some spring scales are calibrated in units of mass, and some balances are calibrated in units of weight.

The ideas of gravitational and inertial mass are difficult to understand without first understanding certain laws of physics, Newton's laws of motion. Let us consider these two quantities in a very descriptive way. Gravitational mass is the kind of mass you think about when you associate mass with the amount of matter in something—with its heaviness. It is the kind of mass we worry about when we are trying to "lose weight". We are really trying to decrease our gravitational mass.

Sir Isaac Newton formulated a law of mechanics which related acceleration (the rate at which an object speeds up) to amount of force (the push or pull on an object). His law gave a proportionality between force F and acceleration a, $F \sim a$. The proportionality constant was a quantity called *inertia*. We now call this quantity *inertial mass*. It is the measure of the opposition that an object has to being speeded up, or accelerated. An object that has a large inertial mass is hard to get going faster and faster. A truck loaded with sand starting off from an intersection accelerates very slowly. The force exerted by the engine is tremendous, but the truck has a very large inertial mass opposing acceleration.

You might have suspected that inertial mass could be the same thing as gravitational mass, since a truck loaded with sand not only has a large inertial mass, but it also has a large gravitational mass. In fact, our experience shows that all objects that have a large inertial mass also have a large gravitational mass. We can show that gravitational mass, M_g, is normally proportional to inertial mass, M_I; $M_g \sim M_I$. But that does not mean they are the same thing. Part of Albert Einstein's great Theory of Relativity did finally establish the equivalence of these two kinds of mass. But at speeds near the speed of light, they become different. They are actually related by the equation

$$M_I = \frac{M_g}{\sqrt{1 - (V/C)^2}}$$

where V is the speed of the object and C is the speed of light, 3×10^8 m/sec. You can see that when V approaches the value of C, we would have the denominator of the right member of the equation approaching zero. That means the right side of the equation approaches a quantity infinitely large. Thus the inertial mass, the opposition to speeding up, would become infinite. From this result we know that no physical object can move as fast or faster than the speed of light.

With a better understanding of the words gravitational mass, inertial mass, and weight, we can be more careful in discussing processes of "weighing", and "comparing masses". We learn to talk about tables of atomic masses instead of about tables of atomic weights. When we use a balance to find that an object has a mass of 12.2 g, we describe that process as comparing an unknown mass with a standard mass, instead of "weighing" an unknown mass.

As Figure 7-7 shows, the simplest balance consists of two equal "arms" pivoted in the center by a knife edge support. To use this balance, an unknown mass is placed in the pan on the right. Then standard masses in units or sub-units are placed in the left pan until the balance indicator points upward. Under this condition, the mass in the left pan is equal to the mass in the right pan. By counting the units of mass in the left pan, the unknown mass is measured.

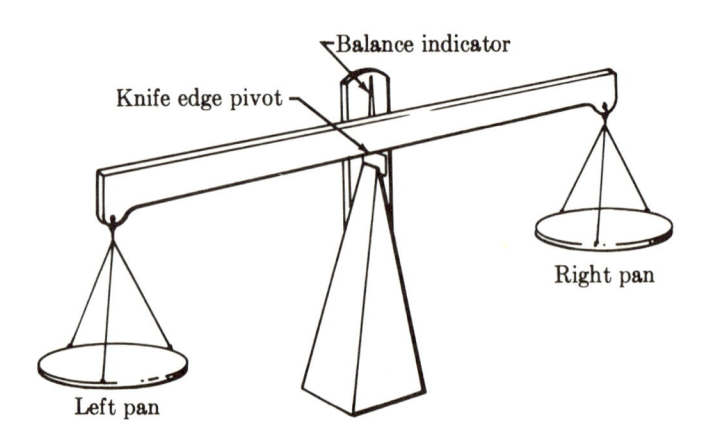

Figure 7-7. *An equal arm balance.*

Let us consider a specific example. A student has an unknown mass which he wants to measure. He places it in one of the pans of his equal arm balance. Then he starts adding standard masses, masses which have been compared with a standard by the manufacturer, to the other

[§ 7-3]   *Simple Measurements of Mass, English and Metric Units*   **115**

pan. He finds that he must add two 100 g masses, four 10 g masses, eight 1 g masses, and three 0.1 g masses. What is the mass of the unknown? The unknown mass must be the sum of the standard masses used to balance the unknown. It would be 248.3 g.

The principle of the equal-arm balance is simple. The force of gravity pulls on each pan with a force proportional to the mass on that pan. If the masses are equal, then the force of gravity on each pan will be the same; thus the torque tending to rotate the balance is zero, and the indicator points upward. Notice that the principle of the balance does not depend upon the size of the force of gravity, but only upon the requirement that gravity not change over the length of the balance. For a source of gravitational field as large as the earth, this requirement is easily met. The fact that the principle does not depend upon the size of the gravitational field means that a balance will work properly here on earth, or on the moon, or on any other planet; a spring scale would give quite different measurements as we moved from one planet to another. The spring scale *does* depend upon the strength of the gravitational field at the location where it is to be used.

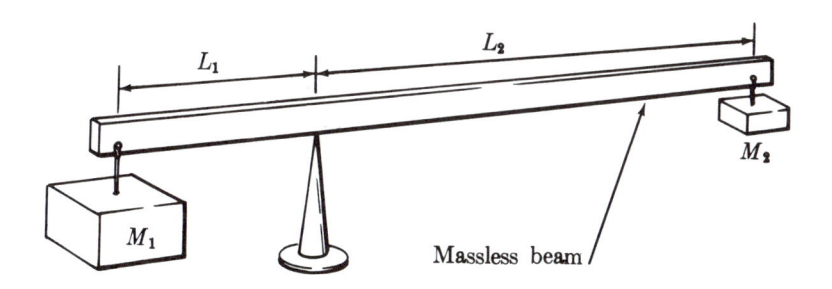

Figure 7-8.   *An unequal arm balance.*

Some balances are constructed using unequal arms. Such instruments are based on the principle that under the condition of balance, the distance from the pivot on one side times the mass on that one side is equal to the distance from the pivot on the other side times the mass on that side. Stated symbolically, using the symbols shown in Figure 7-8, we would say that

$$M_1 L_1 = M_2 L_2$$

Dividing both sides of this equation by L_1, we have

$$M_1 = (L_2/L_1)M_2$$

If we then let side 1 be where we place an unknown mass, and side 2 be where we place the standard masses, we can use the symbol M_u for unknown mass, the symbol M_s for standard mass, L_u as distance from pivot to unknown mass, and L_s as distance from pivot to standard mass. Then the equation becomes

$$M_u = (L_s/L_u)M_s$$

If we construct a balance with $L_s = 10$ cm and $L_u = 1$ cm, then we have $M_u = 10 M_s$. This equation tells us that we need an unknown mass of 10 g to balance a standard mass of only 1 g. We can use standard masses only 1/10 as large to make our measurements. For example, a certain balance has an arm ratio of $L_s/L_u = 20$. What would be the unknown mass when a standard mass of 3.1 g is used? From our equation,

$$M_u = (L_s/L_u)M_s$$

we have

$$M_u = (20)(3.1) = 62 \text{ g}$$

Actually, when such a balance is used, these computations do not have to be made. The balance is calibrated to "read" the correct mass. The common triple-beam balance is constructed in this way, as are many microbalances.

Figure 7-9 shows a triple beam balance, an instrument where the value of L_s is changed as the weights are moved along one of the three beams as a measurement is made. When a weight is added at the right hand end of the beam to increase the capacity of the balance, one must use the ratio $L_s/L_u = 3.31$. Because of that ratio, an increase of 500 g in the balance capacity would require that we add only $500/3.31 = 151.1$ g on that end. These weights are usually available with such balances, and they are marked with the quantity 500 g, even though the actual mass is only 151.1 g.

To use a triple beam balance, one places the unknown mass on the pan. Then the nearest hundred grams is determined by moving the center weight to the notch preceding the one where the pan rises and remains. The measurement is between those two readings of the center beam. Next the back beam weight is moved to the right in precisely the same way as the center weight was, until it is one notch before the one where the pan rises. Finally, the front beam weight is moved to where the balance indicator on the right hand end of the balance oscillates slowly about the center mark. Under these conditions, the balance is set for a reading.

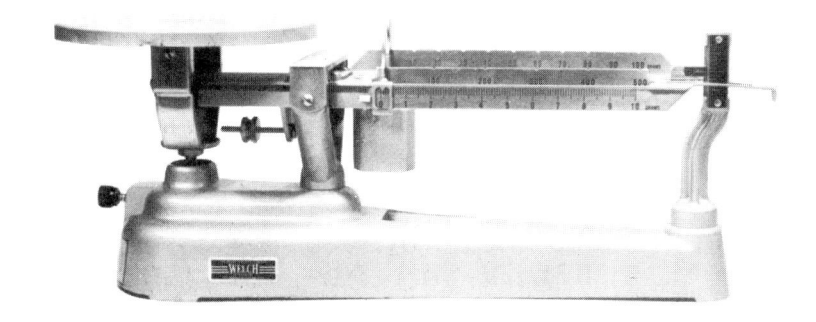

[§ 7-3] *Simple Measurements of Mass, English and Metric Units* 117

Figure 7-9. *A triple beam balance.*

Figure 7-10 shows the settings on a triple beam balance. What is the mass of the unknown? The center beam weight is on the 300 g notch, the back beam on the 40 g notch, and the front beam reads 5.8 g. The balance reading is therefore 345.8 g, the mass of our unknown.

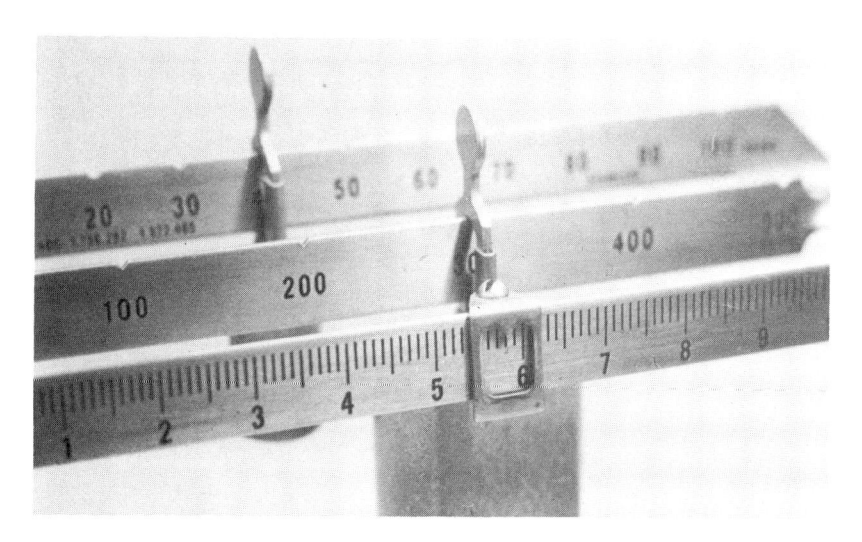

Figure 7-10. *The triple beam balance shows a mass reading of 345.8 g.*

The basic technique of using a balance is the same for all balances, but for a micro-balance, where mishandling can damage the instrument, certain other precautions should be taken. Masses should never be added unless the beam is lifted from the knife edge pivot. Most instruments have a control for permitting this adjustment. The electronic balances also requires special handling. Before using any such instru-

118 Quantitative Aspects of Science and Technology [§7-3]

ment, you should read the operating instructions. With an understanding of the basic principle of the balance, you should be able to understand instructions for using any balance.

PROBLEMS

1. In the following examples, read the triple beam balance and give the mass of the unknown.

a.

b.

c.

[§ 7-3]   Simple Measurements of Mass, English and Metric Units   119
d.

2.   Read the following balance settings, giving the mass of the unknown.
a.

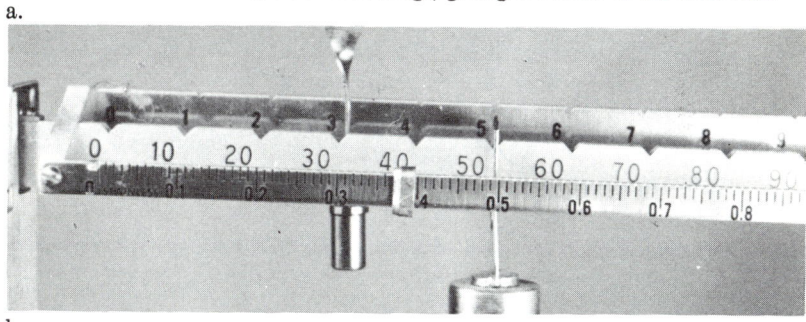

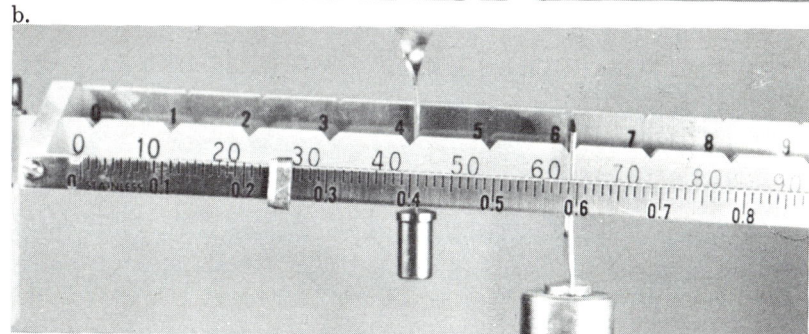

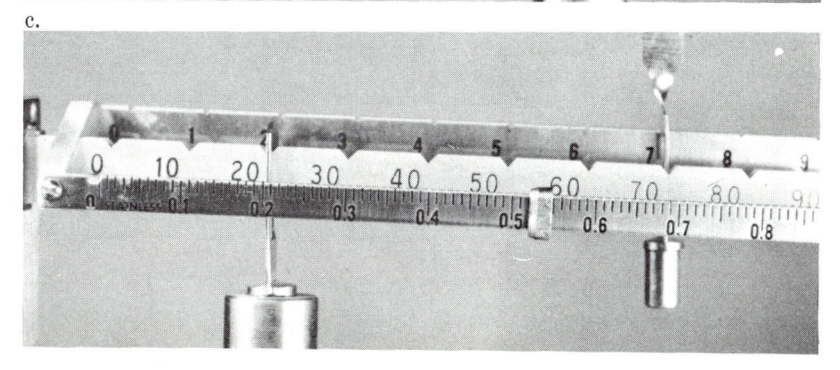

b.

c.

d.

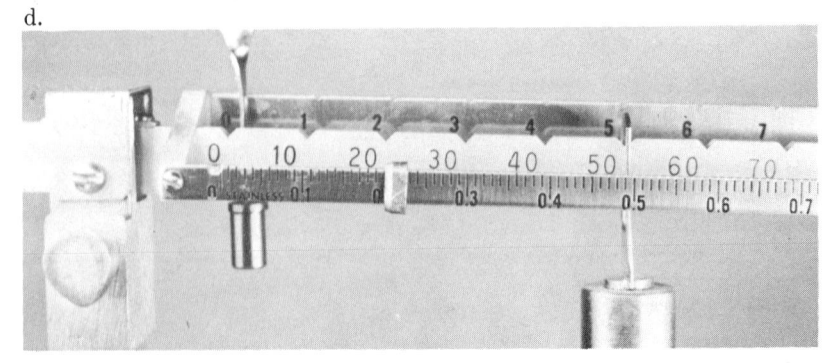

3. A balance has an L_s/L_u ratio of 5.0. What is the mass of an unknown if standard masses of two 100 g masses, seven 10 g masses, four 1 g masses, and nine 0.1 g masses just produce a balance?

4. A balance is constructed so that the pan for the unknown is placed 5 cm from the pivot point and the pan for the standard mass is placed 15 cm from the pivot point. What actual mass would be needed for a standard if a balance is to be achieved with an "unknown" mass of 600 g?

5. A balance has a standard actual mass of 100 g and an "unknown" of 800 g under conditions of a balance. What is the L_s/L_u ratio?

7-4 Measurement of Time Intervals

The concept of time has never been satisfactorily explained by the scientist. All definitions of time are circular, made in terms of the word time itself. We cannot get away from the words period, repetitive, simultaneous, etc., as we try to make a definition. It would seem appropriate, then, to define time intervals in as useful a way as possible, recognizing the imperfections in any definition.

Historically, we have observed objects in the heavens and recorded time intervals in comparison with their motions. We have used the time required for the moon to move around the earth once as about one month. The time it takes the earth to move around the sun once is called a year, and the time required for the earth to rotate once on its axis is called a day. Prior to the 16th century, there was little attempt to accurately subdivide the day. No one really cared too much about smaller time intervals. But since the invention of the pendulum timer by Galileo in the early 1500's, instruments for the subdivision of the day into smaller and smaller units have become common.

The most common device for subdividing the day is the ordinary clock, the kind we are all accustomed to using. It has some kind of inner workings, and two hands which indicate the hour and minute. It is marked off in 12 hours, with 60 minutes per hour. Some clocks have a second hand which divides the minute into 60 parts. Ordinary clocks, or their miniature counterpart, the watch, are seldom able to register time intervals with an error of less than a minute per day. Of course, some fine watches and clocks are able to maintain times correct to within a few seconds per day, but this achievement is still not very remarkable when one considers that an error of three sec per day means the measurement is off 18 min per yr. For some kinds of scientific and technical work, such an error as this cannot be tolerated.

Figure 7-11 shows a stop watch which can register up to 30 min to the nearest 1/10 sec. An examination of the watch face shows that each second is divided into only 5 parts, but the second hand will stop either on a mark, or half-way between two marks. Thus the watch in the figure reads 7.5 sec.

Figure 7-11. *The stop watch reads 7.5 sec.*

For many scientific and technical measurements, time intervals must be specified more accurately and precisely than is possible with an ordinary clock or stop watch. There are devices available today which can measure time intervals with remarkable precision and accuracy. But to use such instruments, the second itself must be better defined.

In 1960, by international agreement, the *second* was defined by the statement that the year 1900 contained 31,556,925.9747 secs. This statement is a definition of the second, and it is a good one. But a standard better than astronomical measurements is now available: the vibrations of an *atomic clock*. The cesium atom radiates energy of a certain vibration rate in the microwave radar region. The *second* is than defined exactly: 1 sec = time for 9,192,631,770 "vibrations" of cesium radiation.

With a good steady definition of the second, the precision and accuracy of a clock can be determined by comparing it with that standard. Because the clock is assumed to be less precise or accurate than the standard, any observed deviations between the ticks of the clock and the ticks of the standard are assumed to be in the clock.

One such clock commonly used in the research and industrial laboratory today is the *electronic interval timer*. Figure 7-12 shows such a timer. These instruments can measure time intervals to within one part in one hundred thousand, considerably better than is possible with a stop watch. Not only is the instrument precise, it is also accurate. It contains a built-in standard that vibrates at a very steady rate. The internal circuitry constantly refers to that internal standard to check its ticking rate.

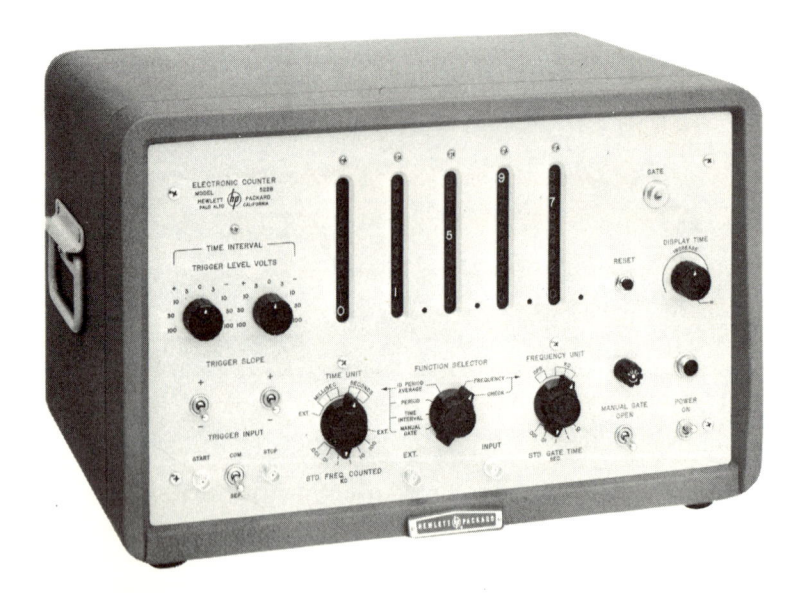

Figure 7-12.   *The electronic interval timer. (Courtesy, Hewlett-Packard Co.)*

There are many kinds of clocks used for measuring time intervals, but the basic principle is the same for all. The clock must have some device to start it ticking. It must have some device for stopping the ticks. It must either have an adjustment so that it can be calibrated against a standard, or it must have a built-in reference standard. Finally, it must have some way of registering the time interval so that we can read it. A clock could be a simple alarm clock, a stop watch, an electronic interval timer, an atomic clock, or even the "ticking off" of particles by a radioactive source. For a radioactive source, the particles leave in a rather irregular and nonuniform way, but over very long periods of time this type of clock can be made precise and accurate in its measurement of time intervals.

PROBLEMS

1. Compute the number of seconds in each of the following time intervals.
    a. 42 days   b. 14 min   c. 4 days, 3 hr, 22 min, 12.3 sec
    d. 4 yr, 87 days, 14 hr, 37 min, 18 sec

2. How long does it take for 400,000,000 vibrations to occur in the cesium atomic clock?

3. In comparing your electronic interval timer built-in standard with a cesium atomic clock at the Bureau of Standards, you find that your internal standard ticks once for 91,926 ticks of the atomic clock. How many times per second does your internal reference standard tick?

7-5 The Calibration of Measuring Instruments

There are two problems associated with any measurement of a physical quantity; these problems relate to precision and accuracy. The precision of a measurement is to be discussed in the next section. Precision involves the fineness of scale divisions on the measuring instrument, and one's ability to read such fine divisions. Accuracy is how well a measuring instrument agrees with a standard. Accuracy is achieved by calibration.

When you use a yardstick, you assume that it actually is one yard long, but it may be too short or too long. The accuracy of any measurement you make with this instrument depends upon whether it has been correctly calibrated. To calibrate an instrument, one must compare it with a reliable standard.

The calibration of a length measuring instrument is usually no problem for the scientist or engineer, unless he needs unusual accuracy.

He can depend upon the manufacturer of instruments to have them properly calibrated, compared with a standard. But for some measurements he may want greater accuracy than a commercial instrument can give. He must then devise some instrument and calibrate it himself, either from a fundamental standard, or in comparison with a Bureau of Standards scale. The standard unit of length, the meter, is now defined as exactly 1,650,763.73 wavelengths of the orange light emitted when an electrical discharge is passed through pure krypton gas of mass number 86. To calibrate a scale, one would allow such light to pass into an interferometer; then by observing circular fringes, waves can be counted as a micrometer screw on the interferometer is turned. A Michelson interferometer is shown in Figure 7-13.

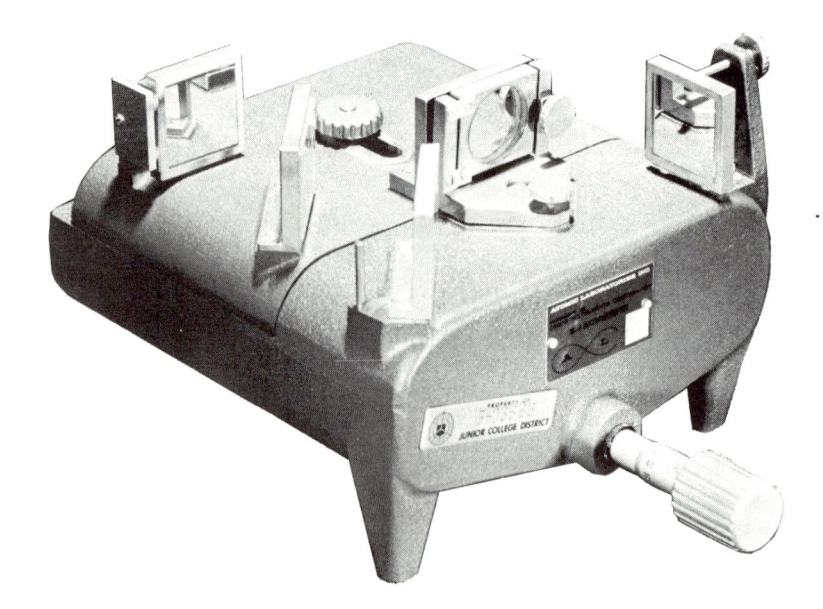

Figure 7-13.   *A Michelson interferometer.*

The problem of calibrating a mass measuring device, a balance, is relatively easy to solve. Standard masses of high accuracy can be puschased. If they are kept clean, they can be used for comparisons directly, or in calibrating an instrument as needed. The calibration of an electronic micro-balance can be checked against a high accuracy standard mass; if the instrument is not accurate, it can be adjusted to remove the error. A mechanical equal-arm balance can be used to compare standard masses directly with unknowns. If a standard mass

[§ 7-5]   *The Calibration of Measuring Instruments*   **125**

is not available, a person can produce his own by placing pure water at 39° F (4°C) into a container having dimensions of exactly 10 cm by 10 cm by 10 cm. This 1000 cm^3 volume of water will have a mass of one kg which is accurate to three parts in one hundred thousand. Of course, the mass of the container must be excluded.

The task of calibrating an instrument for measuring time intervals also depends upon the accuracy desired. For electronic calibration of clocks, the National Bureau of Standards broadcasts time signals which are accurate to two parts in one hundred million. These time signals are broadcast on frequencies of 2.5, 5.0, 10.0, 15.0, 20.0, and 25.0 Mc/sec. With a shortwave radio receiver, these frequencies can be "tuned in". Then the time signals can be synchronized with a clock mechanically or electronically, depending upon the accuracy desired. A person might just set his watch by tuning in on one of these frequencies, listening for the voice announcement which is made every five minutes. The announcement is, for example, "National Bureau of Standards WWV, Fort Collins, Colorado. Next tone begins at fourteen hours forty five minutes, Mountain Standard Time." After a pause of about two seconds, a tone begins with a tick each second heard in the background.

A better, more accurate standard would be the cesium atomic clock discussed in Section 7-4. It is with this type of clock that the WWV broadcasts are synchronized. But to be concerned with this kind of accuracy is unlikely unless the time interval measurements made are to be extremely accurate.

For purposes of ordinary calibration, it is possible to purchase crystals which vibrate at fixed rates. These crystals tick off at rates which are compared with an electronic circuit that can vibrate at various rates. The electronic interval timer usually has such an internal standard with which it can make comparisons to check its ticking rate. For less accurate intervals, a good quality audio or radio frequency oscillator, which has been calibrated at a factory, can be used to provide variable ticking rates.

Regardless of the time interval to be measured, the accuracy of the clock is determined by how it compares with some standard of time. Any clock should be periodically checked against a standard to be sure that it properly measures time intervals, without giving results that are either too large or too small.

The problem of calibration has been discussed so far only in terms of the fundamental physical quantities of length, mass, and time. But the problems of calibration extend to measurements of all physical quantities. Whenever an instrument is designed and constructed to measure some physical quantity, it must be calibrated. That is, its accuracy must be established. For many instruments, the process of calibration must be accomplished continuously. For others, it may be a periodic check against some reference. But it must be done, if the measurements are to mean anything.

Figure 7-14 shows an oscilloscope for which time measurements are determined by distances along the horizontal of the face of the cathode ray tube (like the television picture tube). As the knob settings on the right of the figure indicate, the amount of time per centimeter depends upon where the knob indicator is pointing. But if this oscilloscope is to measure times accurately, the calibration of those sweep settings must be checked with a reference. If they are not correct, adjustments inside the oscilloscope must be made to calibrate it.

Figure 7-14.   *The oscilloscope. (Courtesy, Hewlett-Packard Co.)*

Figure 7-15 shows a digital voltmeter. This instrument has built into it a voltage reference source, a "battery" whose voltage is accurately known. As the voltmeter is used, its electronic circuitry almost continuously checks its own accuracy against the built-in reference. The built-in reference is checked against other standards to make certain of its value.

[§ 7-5]   *The Calibration of Measuring Instruments*   **127**

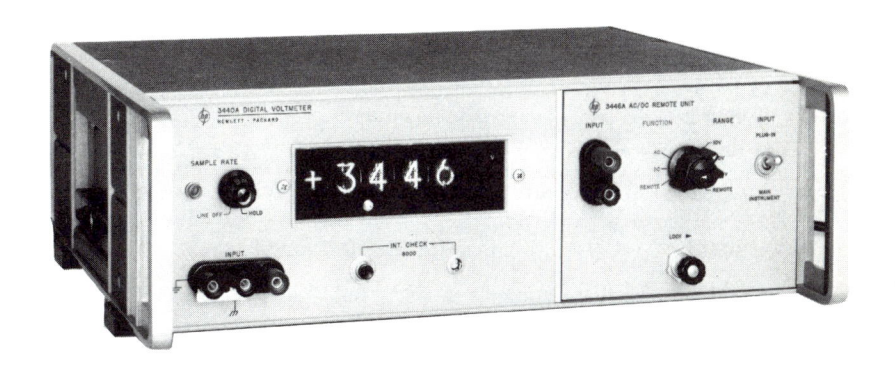

Figure 7-15.   *The digital voltmeter. (Courtesy, Hewlett-Packard Co.)*

As these various examples show, the process of calibration must be carried out for all measuring instruments. Regardless of how much precision an instrument has, its accuracy is entirely dependent upon the care with which it is compared with a standard, and the accuracy of the standard used.

PROBLEMS

1. A scientist wants to make a length measurement with extreme accuracy. He does not have a standard unit of length, but he does have a Michelson interferometer and a source of orange light from krypton 86. He observes circular fringes in his interferometer and determines that 1,651 wavelengths are equivalent to 100.0 turns of his micrometer screw on the interferometer. How much distance does one turn of the micrometer screw represent?

2. An engineer wants to make a mass measurement to an accuracy which is better than his standard masses will permit. He has a bottle which has a very accurate volume of 22.431 cm^3 at a temperature of 39° F. He first balances the empty container as carefully as possible with standard masses. Then he adds pure water to the bottle, filling it exactly.

The room temperature is 39° F. He finds that to get a balance he must add standard masses as follows: two 10 g masses, two 1 g masses, three 0.1 g masses, and five 0.01 g masses. What total mass did he add to balance the water, and how accurate are his standard masses?

3. An audio oscillator that has been calibrated at the factory is guaranteed to be accurate to one part in one thousand. When the oscillator is ticking at 4000 times per sec, a time interval is observed to require 200 of these ticks. What is the length of that interval?

4. A crystal is observed to vibrate at a rate of 20,000 vibrations in each sec, compared with the vibrations of an atomic clock. How long does it take for this crystal to vibrate 528,000,000 times?

5. A digital voltmeter contains a mercury cell as a voltage reference. The cell normally has a voltage of 1.35 v. When the digital voltmeter is compared with a better standard cell having a voltage of 1.1862 v, the voltmeter reads 1.21 v, about 0.02 v too high. What has probably happened to the voltage of the mercury cell?

6. An ammeter is an instrument which measures electrical current. An ammeter constructed by a technician is placed in a circuit where a standard ammeter shows that the current is 30.0 amp. If the constructed ammeter measures linearly, and reads 54.0 in that circuit, what will it read when the current is 5.0 amp?

7-6 Absolute and Relative Uncertainties in Measurements

As was discussed in the previous section, accuracy pertains to the comparison of an instrument with a standard, but precision is related to the fineness of the subdivisions on the instrument and one's ability to read these subdivisions. As an example, let us assume that a meter stick is accurate—that it has been calibrated properly against a good standard. The meter stick has smallest subdivisions of millimeters, one thousand of these marks placed along the stick. A measurement with the meter stick involves counting units and sub-units, but the last digit selected must be a guess. Let us say that a measurement lies between 1.413 and 1.414 m. That last digit, either the third or fourth millimeter mark, must be selected. If it looks as though the measurement is less than half way, we would select the digit 3, but if it is more than half way, we select the 4. Since there are no other marks on the meter stick, we can be sure of no more digits. This measurement, then, could be in error by as much as 1/2 mm, for if the measurement is half way between the third and fourth millimeter marks, we would not know which to

[§ 7-6]   *Absolute and Relative Uncertainties in Measurements*   **129**

select. We could use either digit, recognizing that our error is 1/2 mm anyway.

The example just discussed illustrates a general principle of measurement. For a single measurement, the precision error of the instrument is one half of the smallest main scale division. Provided the instrument is calibrated perfectly, so that no additional errors exist due to poor calibration, and no errors are made in reading the instrument, this is the maximum error.

The sum of all errors for a single measurement is called the *absolute uncertainty* of the measurement. For a meter stick, the absolute uncertainty of a single measurement is 1/2 mm, or 0.0005 m. Using the concept of absolute uncertainty, a measurement made with a meter stick should be written with the stated digits, followed by a plus sign above a minus sign, and then the absolute uncertainty. For example, the measurement 1.413 m would be written (1.4130 ± 0.0005) m. We are then saying that the measurement is somewhere between 1.4125 and 1.4135 m.

When a vernier scale is attached to a measuring instrument, the above error estimate is probably too large. With a vernier scale, the error is reduced to $1/N$ of the smallest main scale division, where N is the number of spaces on the vernier. For the metric scale of a vernier caliper, $N = 10$, therefore the absolute uncertainty is $(1/10)(0.5)$ mm. A measurement might be 4.130 cm. Taking into account the absolute uncertainty, the measurement would be stated (4.130 ± 0.005) cm.

On an ordinary voltmeter, where the needle moves continuously from zero to the full scale value, the same principle applies. The absolute uncertainty is one half the smallest scale division. For example, on a voltmeter reading from zero to 10 v with smallest subdivisions of 0.1 v, the absolute uncertainty would be 0.05 v. A reading might be (4.30 ± 0.05) v. Even on a digital voltmeter, where digits are displayed and we cannot know how far we are between final digits, the instrument will select the closest one; therefore, the absolute uncertainty is still one half of the smallest scale division; in this case it is one half of the step value of the last digit. For a digital voltmeter whose last digit is in hundredths of a volt, the absolute uncertainty would be one half of 0.01, or 0.005 v.

The stop watch discussed in Section 7-4 has smallest subdivision marks of 2/10 sec, but the second hand can stop on those marks or half way between any two of them. Thus the smallest scale division is really 1/10 sec. What is the absolute uncertainty for a time measurement with such a stop watch? One half of the smallest scale division is one half of 1/10, or 0.05 sec. A measurement using this watch, if it is perfectly calibrated, and if synchronization is perfect and without error, could be (8.35±0.05) sec. Of course, if there are other errors present, the absolute uncertainty would be larger.

The preceding measurement with a stop watch suggests that many other sources of error are present when a measurement is made. The quantity one half of the smallest scale division is an *instrument precision error* that is always present. Therefore, under ideal conditions, this error would be the maximum error in a single measurement. But if the edges of an object to be measured with a meter stick were fuzzy, or if one had to synchronize a watch with some event which would depend on his ability to start and stop the watch at correct times, or if the accuracy of an instrument cannot be guaranteed better than 1 per cent or 5 per cent, then these additional errors must be added to one half of the smallest scale division to get the absolute uncertainty.

The concept of absolute uncertainty is useful, and these quantities must be estimated whenever measurements are made in a scientific or engineering laboratory. Very often, the statement of these errors and how they affect some computation containing the measurements can be a critical factor in deciding upon the validity of a theory, or of some engineering design consideration.

Although absolute uncertainties are needed when measurements are made, when measurements are compared with each other the absolute uncertainty is not a good measure of the relative precision of the two measurements. A measurement of the length of a runway at an airport is (3,128±1) m. A pencil has a thickness of (4.4±0.1) mm. The absolute uncertainty in the runway length is 1 m, 10,000 times larger than the absolute uncertainty in the thickness of the pencil. But the measurement of the runway is more precise than the measurement of the pencil. The runway measurement was good to one part in 3,128, but the measurement of the pencil was good only to one part in 44.

The example just discussed suggests that some other quantity be defined to compare the precision of two measurements. This definition has been made, and it is called *relative uncertainty*. Relative uncertainty is found by dividing the absolute uncertainty by the measurement. As an equation, we would write this definition

$$(\text{relative uncertainty}) = \frac{(\text{absolute uncertainty})}{(\text{measurement})}$$

Relative uncertainties can be changed to percentages by multiplying them by 100.

To indicate absolute uncertainty, we use the symbol δX, which is read "delta X". Then relative uncertainty can be written $\delta X/X$. A percentage relative uncertainty could then be $100(\delta X/X)$.

Let us compute these quantities for the length of the runway and the thickness of the pencil mentioned previously. For the runway, the relative uncertainty is $1/3128 = 0.00234$. For the pencil, the relative uncertainty is $0.1/4.4 = 0.0247$. The runway measurement has a smaller relative uncertainty than the pencil measurement. The percentage errors would be 0.234 per cent and 2.47 per cent respectively.

When the scientist or technician wants to make a measurement, he states the measurement plus or minus his best estimate of the absolute uncertainty. Then, when he wants to find out how his measurement error affects some result in which this measurement is included, he must know the relative uncertainty. The methods of analyzing such errors will be discussed in a later chapter of this book.

PROBLEMS

1. A protractor is used to measure the size of an angle. The smallest scale division is one degree, and the measurement is 18 degrees. What is the absolute and relative uncertainty for this measurement?

2. A voltmeter has smallest scale divisions of 0.2 v. What is the relative uncertainty in a reading of 9.6 v?

3. A meter stick is used to measure the width of a table top. What is the relative uncertainty in a measurement of 75 cm?

4. A stop watch has smallest scale divisions of 0.1 sec. The response error for the person starting and stopping the watch is a total of 0.2 sec. If the watch is accurate to within 0.1 per cent, what is the relative uncertainty in a measurement of 5 min 38.7 sec?

5. An ammeter has smallest scale divisions of 0.1 amp. The manufacturer guarantees the instrument to be within an accuracy of 5 per cent of any reading. What is the relative uncertainty in a reading of 12.4 amp?

8 Experimental Data

The science or engineering laboratory is a place where measurements are made. The processes of data collection require specific attention, care being taken not to depend upon one's memory to recall details of the experimental arrangement or of measurements made. Data must be recorded in a way that can be interpreted by other competent people. Some data must be summarized in a presentable form. For measurements where repetition gives variations which are larger than the instrument precision of one half of the smallest scale division, other methods must be used to express the absolute uncertainty in the measurement.

8-1 Fundamental and Derived Quantities

Chapter 7 was concerned with measurements of the fundamental physical quantities length, mass, and time. Mention was also made of other physical quantities like voltage and electric current. These were examples of derived physical quantities.

The fundamental quantities length, mass, and time can be measured with fundamental instruments like the meter stick, the balance, and the clock, respectively. But derived quantities require that two or more fundamental quantities be measured, or else that the measuring instrument be designed to take two or more of these quantities into account, registering the result of that consideration. Although we have dealt with only three fundamental physical quantities, it must be remembered that there is a fourth, electric charge. Even though we have not discussed an instrument for measuring electric charge, there are many instruments which measure derived quantities involving this quantity. Electric current is a derived quantity involving charge and time and is measured by an ammeter. The voltmeter measures the derived quantity voltage, which involves length, mass, time, and electric charge—all four fundamental quantities.

How are derived quantities defined? We saw two examples of such definitions in Chapter 1. The definition of average speed was given by the formula $\overline{V} = D/t$, where $\overline{V}$ is the average speed, D is the distance traveled, and t is the time. An instrument which measured average speed would then be indicating a derived quantity involving the two fundamental quantities length and time. A more complicated derived quantity was pressure, defined by the equation $P = F/A$, where P is pressure, F is force, and A is the area over which the force acts. This derived quantity is defined in terms of area, which is length squared, and force, which is itself a derived quantity. Actually, pressure involves length, time, and mass in a rather complicated way.

Why should we worry about derived quantities? Our only concern is that we may at some time have to calibrate an instrument that measures such a quantity. If we should need to do so, the calibration would have to be done in terms of the fundamental quantities. If we wanted to calibrate a speedometer on an automobile, an instrument which measures the derived quantity speed, we would use the fundamental length standard to lay out a fixed distance along some highway. Then we would use the fundamental time standard to set our clock. By driving the car along this road at some fixed speed, say 60 mph, we would expect to travel one mile in 60 sec. Thus we could calibrate the speedometer.

In recording experimental data, a distinction should be made between derived quantities and fundamental measurements. In many experiments, derived quantities cannot be measured directly anyway; they must be computed from measurements of fundamental quantities.

PROBLEMS

1. Which of the following physical quantities are *not* fundamental?
   a. pressure   b. mass   c. force   d. energy   e. distance
   f. speed   g. electric charge   h. voltage

2. A weight is dropped from rest, falling a distance H in a measured time T. It is observed that the weight speeds up, moving faster and faster uniformly as it falls. Because of this observation, a definition is made of the quantity "acceleration of gravity", symbolized by g. This definition could be $g = 2H/T^2$. Thus g is a derived quantity involving length and time. To make a measurement of g, an engineer measures $T = 2.00$ sec for the weight to fall a distance $H = 64.32$ ft. What would be his measurement of g?

3. The following are derived physical quantities, defined in terms of the fundamental physical quantities length L, mass M, time T, and electric charge Q. In each case determine which of the four fundamental quantities are involved.
   a. $V = L/T$ b. $E_k = (1/2)MV^2$ c. $A = (V_2 - V_1)/T$
   d. $F = MA$ e. $I = Q/T$ f. $P = F/L^2$

8-2 The Experimental Method, Control, and Experimental Variables

We often talk about experimentation, and it seems that the scientist's chief concern is performing experiments. Yet the *process* of experimentation is seldom discussed. Also, it is assumed that there is some fixed method of investigation that has been with us throughout history. Actually, experimental method has evolved over the last five or six hundred years to what is now practiced. Even today, there is no fixed set of rules to follow in performing experiments. Some of the great scientific discoveries have been accidental, and the unusual approach, the method different from an established pattern, has been a critical factor in many great discoveries. But for many kinds of experiments there is a set of conditions which, when present, can lead to discovery.

We should first distinguish *measurement* from *experiment*. Many of the tasks performed by the engineer or technician are simply measurements of derived quantities; they are not experiments. A measurement, like that of the acceleration of gravity, requires that several different fundamental measurements be made. Then these results are used in a defining equation to compute the desired "measurement" of the derived quantity. An experiment involves changing one variable, while holding all but one other variable constant, and watching what it does.

An example of a measurement was discussed in Chapter 7. An engineer was measuring the value of the acceleration of gravity g from a definition, $g = 2H/T^2$. In this laboratory situation, the time T for an object to fall a distance H from rest would be measured. These fundamental measurements would then be used in the defining equation to compute the value of g. The tasks of setting up the apparatus, accurately measuring H, and arranging some device for starting and stopping an accurate and precise timer would all require ingenuity and careful thought. Even then, the engineer might have found that repetition of his measurement gave differences greater than instrument precision errors would have predicted. He may then have repeated the measurement of g many times, finding statistical results for the measurement and for the absolute uncertainty. With all this effort and attention to detail, this process would still not be an experiment. It would be a measurement.

An experiment seeks to find a relationship between two physical quantities. Performing experiments is difficult because a physical situation involves many variables, some perhaps unknown to the experimenter. In the course of an experiment, if more than one variable is permitted to change, it is uncertain which one is producing the changes in the variable being observed.

Let us consider an example of an experiment. We will say that you have learned that a gas can exert a pressure P, have a temperature T, and occupy a volume V. When two of these variables are held constant—not allowed to change—all three appear to remain constant while other variables in the laboratory are changed. Thus we have identified the variables V, T, and P, which characterize the properties of gases. We would like to know the relationship between these variables.

To find the relationship between the variables V, T, and P for gases, we must perform experiments. If we allow the pressure and volume to change, we might observe a change in the temperature, but we cannot tell if it is being caused by the change in the pressure or the change in the volume, or perhaps by both. We must fix one of the variables and not permit it to change. Let us hold the volume V constant and perform an experiment where we change the temperature T and observe changes in the pressure P. In this experiment, we could use a metal container of fixed volume with a gauge attached to read the pressure of some gas placed inside the container before it is sealed. We might use dry helium gas. The container could be immersed first into water boiling at one atmosphere of external pressure to give a temperature of 100° C. Next we could immerse the container in a mixture of ice and water to get a temperature of 0° C. Finally, we could immerse the container in liquid nitrogen at a temperature of $-196°$ C. At each of the above temperatures, we would read the pressure of the helium gas in the container.

In this experiment, we controlled the temperature, held the volume fixed, and watched what happened to the pressure. The variables T and P which were allowed to change are called *experimental* variables. Because we had no direct ability to manipulate the pressure P, its value being dependent upon the temperature, we call the pressure a *dependent* variable. The temperature, which we changed, we call an *independent* variable. When the data from this experiment are analyzed, it is found that pressure P is related to temperature T by the proportion $P \sim (T + 273)$. We have found a relationship between pressure and temperature with volume held fixed. We could design an experiment where the temperature is held fixed, and we would find the relationship between pressure and volume.

In some experiments, a physical system is given some "treatment"; that is, some quantity is present whose influence we want to investigate. In such an experiment, we must have what is called a *control*, and an *experimental* treatment. For example, an experiment in cancer research in Japan involved painting coal tar on one ear of rabbits, but not on the other ear. The unpainted ear was the *control*, the painted ear, the *experimental* treatment. Cancer appeared on the painted ears after an average of 80 days, but not on the unpainted ears. Thus we can assume that the cancer was related to the presence of coal tar.

The preceding examples suggest an experimental method. First some physical system must be identified for investigation. Then, all variables associated with that system must be determined. To find relationships among these variables, all but two must be held constant; then while one of these, the independent variable, is changed, the other, the dependent variable, is observed. A similar procedure must be carried out until all variables have been examined. Then the relationship among the variables can be determined.

PROBLEMS

1. An experiment is designed to find the relationship between the volume and temperature of a gas. The temperature is changed by the experimenter, observations being made on the volume of the gas. What variable must be held fixed? Which is the independent variable? Which is the dependent variable?

2. A researcher in agriculture wants to find out if the growth of a plant is related to the presence of a certain chemical. He places 100 plants in a soil containing the chemical, and 100 more plants in another soil that is identical except that it lacks the chemical. He carefully controls the light, heat, temperature, water, and gases in the air surrounding the plants to be sure that, except for the presence of the chemical, the two plots are identical. Which is the control? Which is the experimental treatment? If he observes after some time that the plants which had been exposed to the chemical have grown an average of three inches higher than the other plants, what might he conclude?

3. An electronics technician performs an experiment in which he holds the electrical resistance R fixed in a simple circuit. He then changes the voltage across the circuit, observing the change in the current. He gets the following data:

V (volts)	I (amperes)
4	2.0
8	4.0
12	6.0
7	3.5

Which are the dependent, and which the independent variables? What is the relationship between these variables?

8-3 *Recording Experimental Data*

Making a derived measurement or performing an experiment each involves the collection of data. Various measurements must be made and recorded. The person making the measurements cannot depend upon his memory to recall some measurement. Also, all variables that might be associated with the system under investigation must be observed. The data must be organized in a form that *any* competent person could analyze. This requirement is the essence of scientific method. An experiment should give results that are completely independent of the person making the observation. Any competent person should be able to examine a set of data and draw the same conclusions from it as any other competent person. It is this objectivity that distinguishes science from other areas of human activity. The laboratory data sheet is the critical step in the performance of any experiment.

The first entry which should be made on a data sheet is the date and time of the experiment. It may not seem that time is a variable of importance, but the investigator does not always know what variables are important in a given situation. Also, the date may turn out to be important later to establish the order in which certain experimental arrangements were tried.

The next entries on the data sheet should be what are called *extraneous variables*, quantities that can be measured easily, that probably remain constant during the experiment anyway, and are assumed to be unrelated to the experiment. These quantities include the room temperature, the barometric pressure, and any other quantitative or qualitative observations which can be made about the experimental surroundings. Then, when the data are being analyzed if one of these variables should become important, it will be available. Without such measurements, an entire set of data could be found useless because of some missing measurement that, unknown to the researcher at the time of the experiment, was a critical factor in the experiment.

Somewhere on the data sheet, a careful description of instruments used and their arrangement should be given. This description would include the accuracy of each instrument, the instrument precision, and some diagram of the relation among the various pieces of apparatus. Someone else examining the data must know how they were acquired.

The basic method of taking data involves the use of tables. The table consists of a table heading, followed by entries. The heading has symbols standing for physical quantities, with the units in which they are measured placed in parentheses. The symbols must be carefully defined. Because an experiment usually involves only two variables, all others being held fixed, the table contains only two columns. The variables held fixed should be identified, and their fixed values recorded next to the table.

When the laboratory problem is one of making a derived measurement, it may not be necessary to use a table, since each of the fundamental quantities measured is used in a formula definition to find the measurement. But sometimes the measurements are repeated a great number of times, so that a statistical technique can be used in determining the "best" measurement and its uncertainty. In that case, the table used could have several variables in the heading, but all entries would be repeated measurements under the same conditions. The fact that they differ slightly would be due to the lack of precision of the instruments used, and not to some change occuring, as with an experimental problem.

Figure 8-1 shows a sample data sheet for an experiment conducted to find the relationship between the electric current I and the voltage V in a circuit having a fixed resistance R. Notice that the data sheet is a record of observations made; it does not include computations or calculations. It is a summary of what was observed, where it was observed, how it was observed, and when it was observed.

For data involving a derived measurement, the entries would appear as shown in Table 8-1. All other quantities would be entered on the data sheet as in an experiment. But for the problem of a derived measurement, the repetition of fundamental measurements requires the use of a table. Actually, it is possible that such a measurement might be only one part of a very complicated experiment.

TABLE 8-1

A set of data involving repetition of measurements of fundamental quantities

Trial	T (seconds)	H (feet)
1	0.995	16.08
2	0.996	16.03
3	0.994	16.00
4	0.998	15.98
5	0.997	16.03
6	0.996	16.06
7	0.998	15.99
8	0.999	16.05
9	0.993	16.04
10	0.993	16.04
11	0.996	16.03
12	0.997	16.07
13	0.994	16.01
14	0.995	16.03
15	0.996	16.02
16	0.995	16.04

T = time for mass M to fall from rest to distance H.
H = distance through which mass M falls from rest.

Date: Dec 15, 1966
Time: Began 1:30 P.M.   Stopped 4:40 P.M.
Place: Saint Louis, Missouri

Extraneous variables: Room temperature 24°C and constant
                        Barometer reading 732 mm Hg and constant

Experimental arrangement:

Circuit diagram

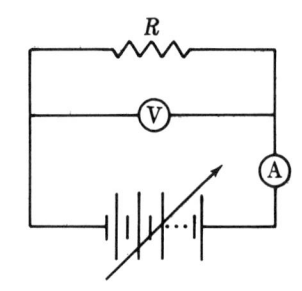

R = standard 20 Ω, 1000 W fixed resistance, accuracy ± 0.001 per cent
V = variable voltage across resistance R
I = variable current through resistance R

Voltmeter: Vacuum tube voltmeter, current drain less than 0.0001 amp, accuracy ± 0.001 per cent
Ammeter: Accuracy ± 0.001 per cent

Experimental Data:

R = 20 Ω, constant

I (amperes)	V (volts)
1.00	20.0
2.00	40.0
2.50	50.0
4.00	80.0
0.50	10.0
1.50	30.0

Figure 8-1.   *A sample data sheet for an electrical experiment.*

[§ 8-4]   *The Mean of a Set of Measurements*   **141**

There are no fixed rules that can be applied to every experimental or measurement situation. In recording data, the best approach is to plan carefully what measurements are to be made, and prepare in advance a neat data sheet. Tables should have headings already labeled. Figures or diagrams of apparatus should be clearly shown. Then, when the experiment or measurement is to be done, there is much less chance that an important measurement will be omitted.

PROBLEMS

1. Figure 8-1 shows a sample data sheet *after* data have been taken. On a sheet of 8½ by 11 paper, carefully show what this data sheet should look like *before* one enters the laboratory to perform the experiment.

2. A gun enthusiast wants to measure the average speed of bullets fired from a rifle. He mounts the rifle solidly, arranges a photoelectronic switch just outside the rifle barrel to start an electronic interval timer when the bullet passes through a narrow beam of light, and he arranges a similar device some distance away through which the bullet will pass to stop the electronic clock. Construct a complete data sheet which *you* would have available when you *began* such a measurement. Include a list of apparatus needed, a table or tables for data, a diagram of the experimental arrangement, etc.

3. An experiment is to be performed to determine the relationship of the length of a simple pendulum to the time required for the pendulum to swing back and forth once (called the *period* of the pendulum). Measurements will be made of various lengths of the pendulum; for each length, the pendulum will be allowed to swing back and forth 50 times in order to get a better measurement of the period. Construct a complete data sheet which you would have ready when you entered the laboratory to perform this experiment.

8-4 The Mean of a Set of Measurements

In most measurements, and in some experiments, the instrument precision of apparatus used is good enough that when measurements are repeated only slightly different results are observed. The simple task of carrying out the measurement produces errors larger than the instrument precision; or the quantity to be measured is uneven to a degree which is greater than the instrument precision error.

If you were to measure the width of a room carefully using a meter stick, you might encounter both of these difficulties. Starting at a certain point at the base of one wall, you would lay out and count meters; then you would count sub-units until you had your measurement. In successively counting meters, you must mark the beginning of the next meter, perhaps with a pencil. Even if the line is drawn very thin, if the room is three or more meters wide, the error introduced could be as large as three or four widths of the pencil line, which is more than a millimeter. If the measurement is repeated from the same point, and a slightly different result is observed, this instrument precision error is probably the cause of that difference.

If the width of a room were measured repeatedly starting from different points along the base of one wall, variations of several millimeters in the width could be observed, depending upon how carefully the walls had been constructed. Which measurement would be called the "width" of the room?

The problem presented above occurs throughout experimental work in science and technology. The precision of instruments is quite good; consequently, because of non-uniformity in the quantity being measured or because of errors resulting from the measurement process, when repeated measurements are made they are found to differ slightly from each other in a somewhat random fashion. Which measurement is "correct", more accurate? Which one should be used, or how can the set of measurements be analyzed to give a "best" measurement?

To decide what to do with repeated measurements, let us examine the set of measurements in Table 8-2. Examining these numbers, we first notice that the smallest is 2.3 and the largest is 2.7. The "actual" or "best" measurement is most likely somewhere between those two values. The difference of these two values, the largest minus the smallest, is called the *range* of the set of measurements.

TABLE 8-2

Trial	L (mm)
1	2.3
2	2.6
3	2.4
4	2.5
5	2.5
6	2.6
7	2.5
8	2.4
9	2.7
10	2.5

The Mean of a Set of Measurements

To decide on a measurement of L, we might select the one that occurs most often, which is called the *mode*. But for most kinds of measurements, the errors are such that they produce what is called a *Gaussian* or *normal distribution*. For such a distribution, the theory of statistics shows that the best estimate of the quantity is the *arithmetic mean*, which we shall call the mean of the set of measurements.

The definition of the *arithmetic mean* of a set of measurements is a mathematical formula. It is the sum of all measurements divided by the number of trials (the number of measurements which have been summed). To write this formula, we need to use a symbol which tells us to form a sum. That symbol is the Greek letter sigma, written Σ. It is placed right before another symbol which represents the quantity to be summed. For example, for the sum of lengths L, we would write ΣL. This is just a short way of saying "find the sum of all measurements of length L". To define the mean with a formula, we can then write

$$\overline{L} = \frac{\Sigma L}{N}$$

Notice that we have used the letter L with a bar over the top to represent the mean. This symbol $\overline{L}$ is read "L bar", or "the mean of L".

To find the mean of the set of measurements of L in Table 8-2, we must first find the sum of the ten L measurements; then we divide that result by the number of measurements, which in this case is ten. Performing that arithmetic, we get

$$\overline{L} = \frac{\Sigma L}{N} = \frac{25.0}{10} = 2.50 \text{ mm}$$

The mean of this set of length measurements is 2.50 mm. We would use this number as our best estimate of the measurement of the quantity L.

We have seen in the above example how the arithmetic mean of a set of numbers can give a measure which is the best estimate of the value of a quantity where repeated measurements give variations. To decide whether or not to make many measurements of a quantity, we need only make a few initial measurements. If the instrument precision error is not large, and the quantity being measured is uniform, we may get the same measurement each time. In this case, no statistical analysis is needed. But if a few measurements show variations, we would want to take a larger sample, finding the mean as our best estimate of the quantity being measured.

Although we have a way of finding the best estimate of the value of some quantity where variations are observed with a certain instrument, we have not discussed how we would estimate the uncertainty in that mean value. What would be its absolute uncertainty? This question is answered in the following three sections.

PROBLEMS

1. In a practice session, a bowler has the following scores: 148; 176; 190; 158; 170; 175. What is the arithmetic mean of these six scores?

2. A technician is performing a measurement involving the resonant frequency of a tuned circuit. The instrument precision is good enough that repetitions give variations. When he repeats the experiment, he gets the following frequencies: 8.40×10^4; 8.37×10^4; 8.49×10^4; 8.41×10^4; 8.45×10^4; 8.43×10^4. What is his best estimate for the resonant frequency of this tuned circuit?

3. A group of five engineers are making measurements that are strongly dependent upon the atmospheric pressure. To get the best estimate of the atmospheric pressure, each of the five makes an independent measurement, not telling the others of his observation. When they have all performed the measurement, the results are as follows: 74.32 cm; 74.38 cm; 74.36 cm; 74.35 cm; 74.37 cm. What is their best estimate for the atmospheric pressure (in cm of mercury)?

4. Find the best estimate of each quantity in the following sets of measurements.

   a. Trial I (amperes)

1	8.32
2	8.33
3	8.25
4	8.30
5	8.36
6	8.27

   b. Trial V (voltage)

1	107.4
2	107.9
3	107.6
4	106.9
5	107.7
6	107.1

8-5 The Mean Deviation of a Set of Measurements

In the preceding section, we used the arithmetic mean to compute the best estimate of a quantity where repeated measurements gave variations larger than the instrument precision error. But the problem of estimating the uncertainty in this best estimate had not been resolved. We must find some way of estimating the absolute uncertainty in the arithmetic mean.

[§ 8-5]   *The Mean Deviation of a Set of Measurements*   **145**

To further point out this problem, Table 8-3 shows two sets of measurements having the same mean, the same range, and the same number of measurements. The mean in each case is $\overline{L} = 4.5$ mm, and the range is 4.8 mm − 4.2 mm = 0.6 mm. But a careful examination of each set shows that the measurements are more "spread out" in one set than in the other. In Table 8-3(a), there are five measurements at $L = 4.5$ mm, but there are only three such measurements in Table 8-3(b). Our confidence in the mean as a best estimate of L is greater for one of these sets than for the other. The larger the number of measurements close to the mean, the more confidence we have in that mean. Thus, we need a measure of the "spread" of measurements with respect to the mean.

TABLE 8-3

	(a)			(b)	
Trial	L (mm)		Trial	L (mm)	
1	4.2		1	4.5	
2	4.5		2	4.6	
3	4.6		3	4.8	
4	4.3		4	4.2	
5	4.4		5	4.3	
6	4.8		6	4.7	
7	4.5		7	4.4	
8	4.4		8	4.7	
9	4.7		9	4.5	
10	4.3		10	4.5	
11	4.5		11	4.3	
12	4.7		12	4.6	
13	4.5		13	4.3	
14	4.4		14	4.4	
15	4.6		15	4.6	
16	4.5		16	4.4	
$N = 17$	4.6		$N = 17$	4.7	
	$\Sigma L = 76.5$			$\Sigma L = 76.5$	
$\overline{L} = 76.5/17 = 4.5$ mm			$\overline{L} = 76.5/17 = 4.5$ mm		

Table 8-4 shows even more clearly how the spread of a set of measurements is related to the confidence we have in the best estimate of the quantity, and in the estimate of the absolute uncertainty in the mean value. Notice that the first set has all measurements identical, but the other distributions have greater and greater spread. Yet the mean, 50, is the same in each set.

TABLE 8-4

Trial	Set A L (mm)	Set B L (mm)	Set C L (mm)	Set D L (mm)	Set E L (mm)
1	50	52	54	56	60
2	50	51	52	53	55
3	50	50	50	50	50
4	50	49	48	47	45
N = 5	50	48	46	44	40
$\Sigma L =$	250	250	250	250	250

$\overline{L} = 50$ for each set

In examining Table 8-4, we might be tempted to use the range as a measure of the spread of the distribution. The range does increase from zero in Set A to 20 in Set E, but in Table 8-3, where the two sets had the same range, there still was a greater spread in one set of measurements than in the other. We need some other measure.

One such measure of spread is called the *mean deviation*. It is the average amount that the measurements differ from the mean. For example, in Set B of Table 8-4, the difference of the mean and 52 is 2, of the mean and 51 is 1, of the mean and 50 is 0, of the mean and 49 is 1, and of the mean and 48 is 2. The sum of these deviations is 6. Thus the average, or mean deviation is $6/5 = 1.2$. Notice that we have considered only positive differences, taking the larger number from the smaller. Because we are interested in the average amount that the measurements deviate from the mean, we do not care about the sign.

To make a mathematical definition of the *mean deviation*, we must state our definition so that the signs of the differences are always positive. If we do not, the sum of the differences will be zero. To make all differences positive, we define *deviation* as the absolute value of the difference of the mean and each measurement, written $|\overline{X} - X|$. The definition of the mean deviation $\overline{d}$ is then given by the following formula:

$$\overline{d} = \frac{\Sigma |\overline{X} - X|}{N}$$

where $\overline{X}$ is the mean, X is one of the measurements, and N is the number of measurements.

Using the above definition of the mean deviation, we shall compute $\bar{d}$ for Set C of the data in Table 8-4. Let us arrange the data in a convenient form. We will show the column of measurements; then next to that column we show the deviation $|\bar{L} - L|$. The sum of these deviations divided by N is then the mean deviation $\bar{d}$.

TABLE 8-5

Trial	L (mm)	Deviation = $|\bar{L} - L|$
1	54	4
2	52	2
3	50	0
4	48	2
5	46	4
		12mm

$\bar{d} = 12/5 = 2.4$ mm

Although we have no basis for doing so, we might use the mean deviation as a measure of the absolute uncertainty in our measurement. It is certainly larger than the instrument precision, and it is better than no error estimate at all. If we do use the mean deviation as an estimate of the absolute uncertainty, our best estimate of L in Table 8-5 would be $L = (50.0 \pm 2.4)$ mm.

The use of the mean deviation as an error estimate is better than having none at all, but there is no sound basis for suggesting that this estimate of absolute uncertainty is correct, or even reasonable. Some other measure of spread that has some foundation in statistical theory is needed. We will see in the next sections that such a measure exists and is more useful in specifying uncertainties than the mean deviation.

PROBLEMS

1. In each of the following examples, find the deviation of the measurement from the mean.
   a. $\bar{X} = 8, X = 12$ b. $\bar{L} = 54.8$ mm, $L = 51.3$ mm
   c. $\bar{V} = 112$ v, $V = 121$ v   d. $\bar{Y} = 8.0, Y = 8.6$

2. For the measurements of Table 8-4, find
   a. the mean deviation for each set;
   b. an estimate of the absolute uncertainty for each set; and
   c. the relative uncertainty for each set.

3. For the measurements of Table 8-3, compare the mean deviations and estimate the absolute uncertainty of $\overline{L}$ for each case. Compute the relative uncertainty.

4. If we did not use the absolute value of the difference of $\overline{X}$ and X in our definition of mean deviation, we would have some negative differences. Find the sum of differences $\overline{X} - X$ for each set of data in Table 8-4. Why do we use the absolute value of these differences?

5. A scientist makes the following measurements of the electric current in a circuit: 9.82 ma (milliamperes); 9.80 ma; 9.87 ma; 9.86 ma; 9.83 ma. What is his best estimate of the current, of its absolute uncertainty, and the relative uncertainty?

8-6 The Standard Deviation of a Set of Measurements

Even though the *mean deviation* discussed in Section 8-5 can be used as an estimate of the absolute uncertainty in a measurement, there is no basis upon which that selection of the mean deviation can be made. The statistician has another measure, the *standard deviation*, for which some meaning can be assigned.

The standard deviation formula is

$$S = \sqrt{\frac{\Sigma |\overline{X} - X|^2}{N - 1}}$$

Notice that we still use deviations from the mean, but they are squared before finding the sum. Let us calculate the standard deviation for a set of measurements; then we can use those results to discuss the meaning of standard deviation.

TABLE 8-6

L (mm)	$|\overline{L} - L|$	$|\overline{L} - L|^2$	L (mm)	$|\overline{L} - L|$	$|\overline{L} - L|^2$
70	20	400	52	2	4
50	0	0	54	4	16
57	7	49	43	7	49
30	20	400	32	18	324
65	15	225	59	9	81
34	16	256	44	6	36
46	4	16	38	12	144
56	6	36	48	2	4
37	13	169	51	1	1
62	12	144	60	10	100
49	1	1	68	18	324

[§ 8-6] The Standard Deviation of a Set of Measurements   149

TABLE 8-6—Continued

L (mm)	$\lvert \overline{L} - L \rvert$	$\lvert \overline{L} - L \rvert^2$	L (mm)	$\lvert \overline{L} - L \rvert$	$\lvert \overline{L} - L \rvert^2$
50	0	0	59	9	81
55	5	25	46	4	16
53	3	9	75	25	625
25	25	625	45	5	25
45	5	25	52	2	4
36	14	196	67	17	289
63	13	169	51	1	1
52	2	4	35	15	225
48	2	4	50	0	0
39	11	121	54	4	16
58	8	64	36	14	196
61	11	121	53	3	9
41	9	81	63	13	169
54	4	16	37	13	169
33	17	289	52	2	4
38	12	144	62	12	144
66	16	256	42	8	64
56	6	36	49	1	1
46	4	16	58	8	64
47	3	9	40	10	100
53	3	9	47	3	9
44	6	36	57	7	49
64	14	196	60	10	100
42	8	64	48	2	4
51	1	1	61	11	121
54	4	16	45	5	25
40	10	100	39	11	121
58	8	64	56	6	36
64	14	196	59	9	81
50	0	0	41	9	81
47	3	9	43	7	49
31	19	361	55	5	25
69	19	361	46	4	16
56	6	36	44	6	36
43	7	49			
49	1	1			
41	9	81			
57	7	49			
48	2	4			
44	6	36			
55	5	25			
42	8	64			

$\Sigma L = 4900 \qquad \Sigma \lvert \overline{L} - L \rvert^2 = 9702$

$\overline{L} = \Sigma L / N = 4900/98 = 50.0$

$S = \sqrt{\dfrac{\Sigma \lvert \overline{L} - L \rvert^2}{N - 1}} = \sqrt{9702/97}$

$ = \sqrt{100.02} = 10.0$

$\overline{L} = 50.0 \qquad S = 10.0$

Table 8-6 contains a set of 98 measurements of a length L, measured in millimeters. The standard deviation formula indicates that we must first find the mean, $\bar{L}$, for the set of measurements. This has been done by finding the sum of all 98 measurements and dividing by the number of measurements. The mean is 50.0. When the mean is known, the absolute value of the difference of each measurement and the mean is found. These deviations, $|\bar{L} - L|$, make up the second column in the table. The formula for standard deviation requires that we square each of these deviations and find their sum. This has been done in the third column, headed $|\bar{L} - L|^2$. At the bottom of this table we have summed the squared deviations to get 9702. This result is then used in the formula to compute a standard deviation of $S = 10.0$. We have analyzed these 98 measurements to get a mean of 50.0 and a standard deviation of 10.0. Now what does all this mean?

In the theory of statistics, if a distribution is Gaussian it can be shown that 68.26 per cent of the measurements will be from $\bar{L} - 1S$ to $\bar{L} + 1S$. That is, the chance of a measurement being within that range is 68.26 per cent. In the distribution above, 68.26 per cent of the measurements should be between $50 - 10 = 40$ and $50 + 10 = 60$. Figure 8-2 shows these measurements, in order, from 25 to 70, with tally marks indicating the number of each size. Notice that we split the number falling at 30, 40, 60, and 70, so that half are counted above, and half below those borderline measurements. Between 40 and 50, we have 66 measurements (67.3 per cent).

Statistical theory also predicts that 95.46 per cent of the measurements should be between $\bar{L} - 2S$ and $\bar{L} + 2S$. For our set of measurements in Figure 8-2 we have 95 measurements (97 per cent) between 30 and 70. Statistical theory also predicts that 99.74 per cent of the measurements should be between the limits $\bar{L} - 3S$ to $\bar{L} + 3S$. For our distribution, all of the measurements are between those limits, 20 and 80.

Figure 8-3 shows a graph of our 98 measurements. The graph was constructed by grouping the measurements listed in Figure 8-2 into intervals of five each, placing that amount at the midpoint of each interval. Those coordinates are circled points through which the smooth curve has been drawn. The result is a close approximation to the Gaussian normal distribution curve. The rectangles "under" the curve each represent one measurement; therefore, by counting rectangles we could determine the number of measurements in a given interval.

What, then, does the standard deviation mean? If the distribution of measurements is Gaussian, the standard deviation gives the limits for the probability of making a certain measurement. Figure 8-4 illustrates this point. In Figure 8-4(a), the shaded area indicates that

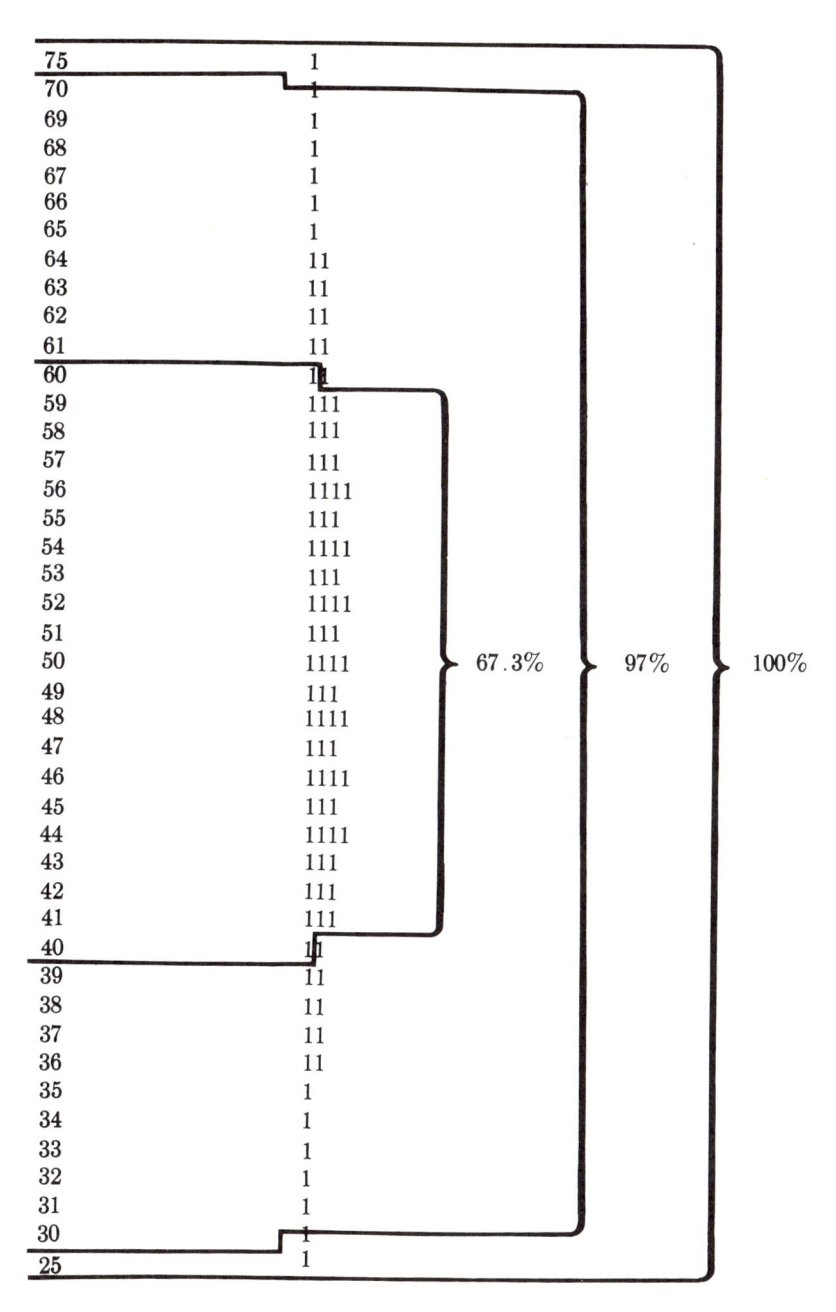

Measurement	Tallies
75	1
70	1
69	1
68	1
67	1
66	1
65	1
64	11
63	11
62	11
61	11
60	11
59	111
58	111
57	111
56	1111
55	111
54	1111
53	111
52	1111
51	111
50	1111
49	111
48	1111
47	111
46	1111
45	111
44	1111
43	111
42	111
41	111
40	11
39	11
38	11
37	11
36	11
35	1
34	1
33	1
32	1
31	1
30	1
25	1

Braces: 67.3%, 97%, 100%

Figure 8-2. *A distribution of measurements.*

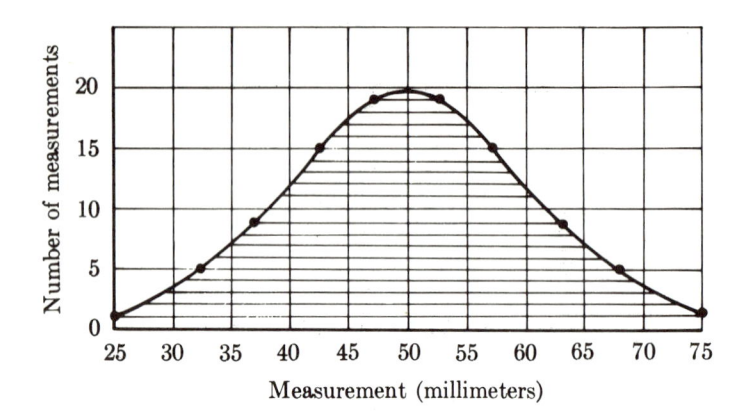

Figure 8-3. *A graph of a normal distribution.*

68.26 per cent of all measurements should be found within plus or minus one standard deviation of the mean. The shaded area of Figure 8-4(b) shows how 95.46 per cent of all measurements should be within plus or minus two standard deviations of the mean. Finally, Figure 8-4(c), being plus or minus three standard deviations from the mean, should contain 99.74 per cent of all measurements.

To say that measurements should fall into these percentage categories really tells us something about the probability of making a measurement of a certain size. If, instead of making only 98 measurements as we did in Table 8-6, we had an extremely large set of measurements, we would expect this distribution to approach a Gaussian distribution as the number of measurements increases. As the percentages showed, our 98 measurements approached the Gaussian distribution closely. The theory of statistics indicates that for the kinds of errors we are discussing, the mean of a very large set of measurements would approach the "true" mean value of the measurement. Because we can never make such a large number of measurements, our mean may not be the same as this "true" value, but we can estimate the error in our mean.

For a very large number of measurements, the mean and standard deviation would be called the *actual* or *real* values of those statistical measures. But since we never have such a large number of measurements, our formulas specify our *estimates* of these unknown real quantities. Let us assume for a moment that we know the real values of the standard deviation and mean for a set of measurements. Actually, the data of Table 8-6 was constructed from an ideal distribution having a

[§ 8-6]   *The Standard Deviation of a Set of Measurements*   153

mean of $\overline{L} = 50.0$ and a standard deviation of $S = 10.0$. If a single measurement of the quantity L were to be made, we know that the chance of that measurement falling between $\overline{L} - 1S$ and $\overline{L} + 1S$, that is between 40 and 60, is 0.6826. About 68 times out of 100, the measure-

(a)

(b)

(c)

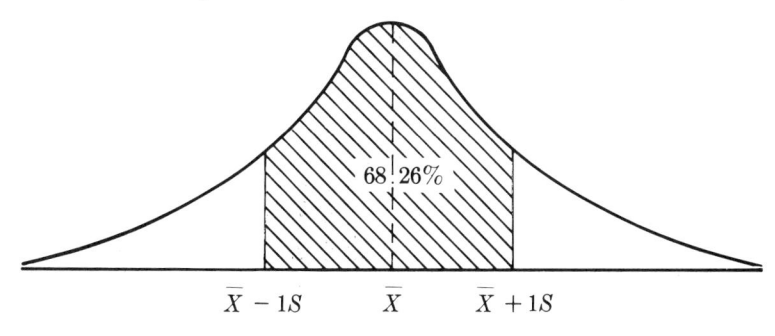

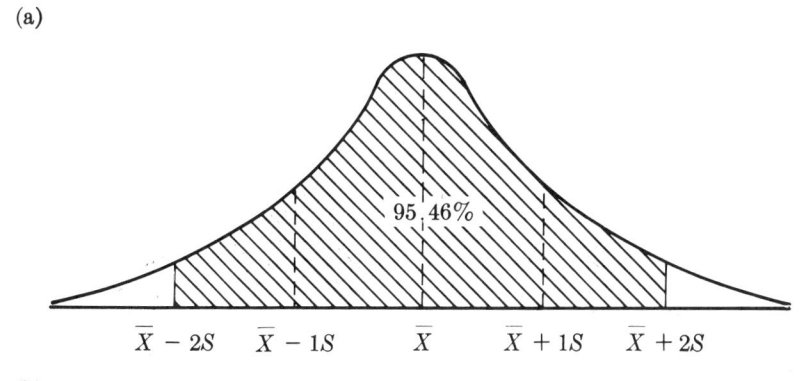

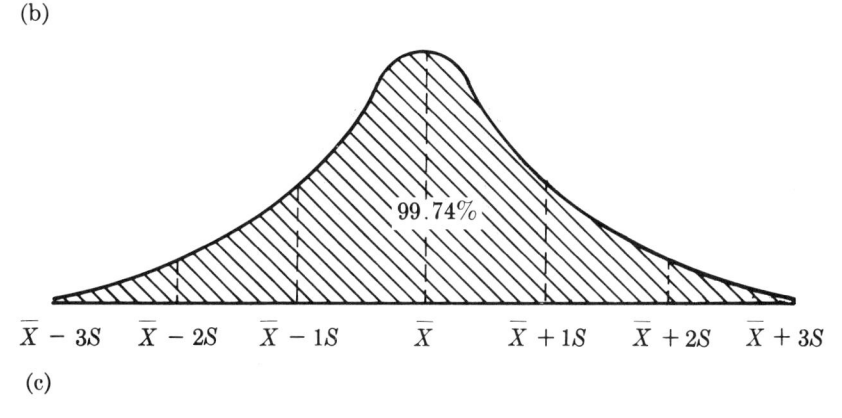

Figure 8-4.   (a) Within $\pm 1S$ of $\overline{X}$.
　　　　　　　　(b) Within $\pm 2S$ of $\overline{X}$.
　　　　　　　　(c) Within $\pm 3S$ of $\overline{X}$.

ment would fall between those limits. About 95 times out of 100, the measurement would fall between $\overline{L} - 2S$ and $\overline{L} + 2S$, that is between 30 and 70. As Figure 8-4(c) shows, almost all measurements (99.74 per cent) can be expected to fall between $\overline{L} - 3S$ and $\overline{L} + 3S$, between 20 and 80 in our set of measurements.

The meaning of standard deviation involves the probability of making a single measurement within a certain interval. But in a practical situation, we cannot know what the standard deviation is unless we make an extremely large number of measurements. What use, then, can we make of the standard deviation?

Actually, we use our formula for standard deviation to *estimate* the real standard deviation. We could not compute the correct result unless we had a very large number of measurements. But in this text, we will assume that our estimate of the standard deviation found with our formula is sufficiently close to the real value that it can be used. However, the mean we normally compute is also only an estimate of the "true" mean. What we want is to be able to use our standard deviation to determine the uncertainty in the mean. This could be the absolute uncertainty in our measurement.

If five different technicians make five different sets of measurements, when they analyze their data they find that they get different means and different standard deviations. If we imagine that an extremely large number of sets of five measurements are taken, and means found for each set, this distribution of means also is a Gaussian distribution and has its own standard deviation. We call this quantity the *standard deviation of the mean*. The statistician has a formula for the standard deviation of the mean. It is

$$S_{\overline{X}} = S/\sqrt{N}$$

where $S_{\overline{X}}$ is the standard deviation of the mean, S is the standard deviation, and N is the number of measurements.

The standard deviation of the mean has the same general interpretation for means as the standard deviation had for single measurements. The probability of a mean for a sample being between $\overline{X} - 1S_{\overline{X}}$ and $\overline{X} + 1S_{\overline{X}}$ is 0.6826, between $\overline{X} - 2S_{\overline{X}}$ and $\overline{X} + 2S_{\overline{X}}$ is 0.9546, and between $\overline{X} - 3S_{\overline{X}}$ and $\overline{X} + 3S_{\overline{X}}$ is 0.9974. Because we do not know the true value of the mean, but only an estimate of it, we must assume that our mean is somewhere in one of these intervals. But our standard deviation estimate allows us to get a close approximation to the standard deviation of the mean using the formula for $S_{\overline{X}}$, therefore we can use the probabilities just discussed to specify how close the real mean is to our computed mean.

As an example, let us select a sample of 10 measurements from the set of 98 measurements in Table 8-6. To make these measurements random, let us select the fifth, then each succeeding tenth measurement. Our sample of ten measurements is shown in Table 8-7. Our mean

TABLE 8-7

L (mm)	$|\bar{L} - L|$	$|\bar{L} - L|^2$
65	14	196
32	19	361
53	2	4
41	10	100
64	13	169
69	18	324
59	8	64
36	15	225
47	4	16
43	8	64

$\Sigma L = 509 \qquad \Sigma|\bar{L} - L|^2 = 1523$

$\bar{L} = \dfrac{\Sigma L}{N} = \dfrac{509}{10} = 50.9$

$\bar{L} = 51$

is 50.9, which we round off to 51 to get deviations. We already know that the larger distribution has an actual mean of 50, thus our estimate is close. Computing the standard deviation S, we get

$$S = \sqrt{1523/9} = \sqrt{169} = 13$$

The actual standard deviation is 10.0, but this value of 13 is not greatly in error. Now let us find the standard deviation of the mean. We get

$$S_{\bar{x}} = S/\sqrt{N} = 13/\sqrt{10} = 13/3.16 = 4.1$$

This result means that with samples of ten measurements, we could expect the true mean to be within a range of $50.9 - 4.1 = 46.8$ to $50.9 + 4.1 = 55.0$ with a probability of 0.6826. We would expect the true value to be within two standard deviations of the mean, from 41.7 to 59.1, with a probability of 0.9546, and within three standard deviations of the mean, from 38.6 to 63.2, with a probability of 0.9974. These ranges tell us how close our computed mean is to the actual mean of some measurement. Thus, the standard deviation of the mean is a measure of the absolute uncertainty for a mean of a set of measurements.

In the example we just worked, the mean of ten measurements, 50.9, was only 0.9 larger than the true mean of 50.0, yet 68.26 times out of 100 our mean would be within plus or minus one $S_{\bar{x}}$ of the real mean. That is, the probability of our mean of 50.9 being within the interval of 46.8 to 55.0 was better than a fifty-fifty chance, actually 0.6826. As a measure of absolute uncertainty, the standard deviation of the mean is somewhat large. In the next section we shall examine another measure of the absolute uncertainty which depends upon the quantities discussed here, but is more often used.

PROBLEMS

1. Using the first ten measurements in Table 8-6, find the mean $\bar{L}$; then, by finding values of $|\bar{L} - L|$ and $|\bar{L} - L|^2$, compute the standard deviation, S, and the standard deviation of the mean, $S_{\bar{x}}$. How much does your mean differ from the actual mean? Within what limits would you expect to find the actual mean with a probability of 0.6826? Within what limits would you expect to make a single measurement of L with a probability of 0.6826?

2. A technician makes a set of 36 measurements of electrical potential difference. When he analyzes these data statistically, he gets a mean of 43.1 v, and a standard deviation of 1.2 v. Within what limits would he expect to make a single measurement 95.46 times out of 100? What would be the standard deviation of the mean? With a probability of 0.6826, within what limits would he expect to find the actual mean voltage?

3. Table 8-1 shows 16 measurements of the time T taken for an object to fall through a distance H. By finding the mean, the standard deviation, and the standard deviation of the mean for both T and H, give, as an estimate of absolute uncertainty in each case, a quantity which gives a range in which the real mean would fall with a probability of 0.6826. What would be the relative uncertainties for H and T?

8-7 Probable Error

The interpretation of standard deviation in the previous section was in terms of probabilities of finding a measurement or a mean of a set of measurements within a certain interval. To get a more useful estimate of the absolute uncertainty of a measurement, or of a mean of measurements, we shall now consider a quantity which gives a

[§ 8-7]   Probable Error   157

probability of 0.5, a fifty-fifty chance, for making a measurement within a certain interval. The size of this interval is determined by what is called *probable error*.

Probable error, symbolized by PE, is somewhat like the standard deviation in that $\overline{X} - 1PE$ to $\overline{X} + 1PE$ contains 50 per cent of all measurements. The probability of making a single measurement within plus or minus one PE of the mean is 0.5, a fifty-fifty chance. The probable error formula is determined by finding 50 per cent of the "area" under the Gaussian distribution curve as shown in Figure 8-5. When this process is accomplished using advanced mathematics, the result is the formula $PE = 0.67S$ for a single measurement. For the probable error of the mean, we must have $PE_{\overline{X}} = 0.67S_{\overline{X}}$.

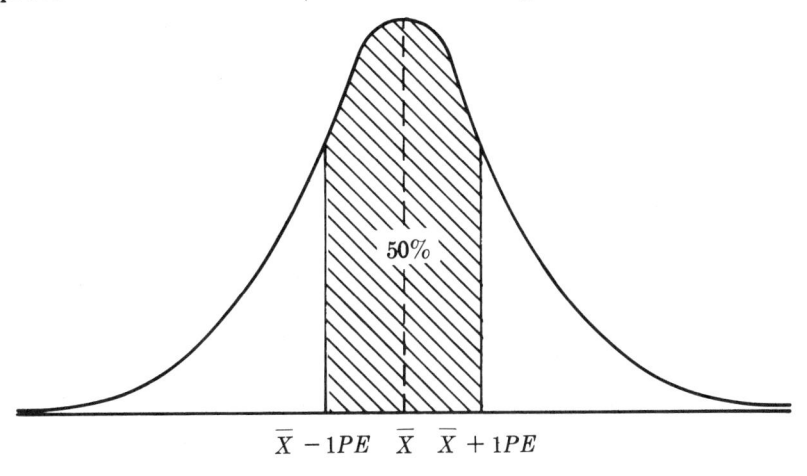

Figure 8-5.   *Within $\pm 1PE$ are 50 per cent of all measurements.*

From the data of Table 8-7 we computed a mean of 50.9, a standard deviation of 13.0, and a standard deviation of the mean of 4.1. The probable error would then be

$$PE = 0.67S = (0.67)(13.0) = 8.7$$

The probable error of the mean would be

$$PE_{\overline{X}} = 0.67S_{\overline{X}} = (0.67)(4.1) = 2.75$$

which we shall round off to 2.8.

Now that we have the probable error and the probable error of the mean, what do these quantities mean? The probable error means that there is a fifty-fifty chance that a single measurement of L will be between $50.9 - 8.7 = 42.2$ and $50.9 + 8.7 = 59.6$. The probable error of the mean has a similar meaning. There is a fifty-fifty chance that another set of ten measurements will have a mean between $50.9 - 2.8 = 48.1$ and $50.9 + 2.8 = 53.7$. More importantly, the probable error of the mean tells us there is a fifty-fifty chance that the actual mean is within plus or minus 2.8 mm of our computed mean. Stated differently, we could say for the actual mean $\bar{L}$,

$$\bar{L} = (50.9 \pm 2.8) \text{ mm}$$

We can use this $\bar{L}$ as the actual mean, with the probable error of the mean as our estimate of the absolute uncertainty in $\bar{L}$.

For a thorough understanding of these various statistical quantities, we must understand the theory of statistical sampling. Each statistical measure has certain limitations, and its use is based on certain assumptions. But it is better to use the mean and probable error of the mean for the measurement and uncertainty in a measurement than to take only a single measurement and be completely ignorant about errors.

PROBLEMS

1. For each of the three problems at the end of Section 8-6 find the probable error and the probable error of the mean. What is the absolute uncertainty in each of the various quantities mentioned in those problems?

2. Problem 4 at the end of Section 8-4 has two sets of measurements. Using the probable error of the mean as the absolute uncertainty, find the relative uncertainty for the voltage and for the current.

9 Graphical Analysis of Experimental Data

The set of data resulting from an experiment is often difficult to analyze numerically. Relationships are sometimes complicated enough that much trial and error manipulation would be required to find an appropriate equation. The methods of graphical analysis are quick and effective for discovering any of a large class of relationships. Even where graphical analysis does not reveal a simple equation for the relationship between experimental variables, the graph itself expresses relationships as well, and in some cases better, than an equation. One need only scan a scientific or technical paper to see how extensively graphs are used to present relationships among experimental variables. The mathematician uses graphs of equations to better visualize certain problems that occur. All equations involving two unknowns can be graphed in a plane, that is on a flat sheet of paper. Each equation has its own peculiar "curve". The ability to read a graph, to interpret certain characteristics of it, and to construct graphs is essential in studying science or technology.

9-1 Coordinate Axes

What is a graph? How do we make one? A graph is a visual representation of numbers. When you learned elementary algebra you began learning about numbers using a "number line". As shown in Figure 9-1(a), numbers are represented by points on the number line, larger numbers being farther to the right than smaller numbers. Negative numbers are placed to the left of zero, the negative numbers having the largest absolute value being farthest left. Each point on the number line is a graph of the number the point represents. The graph of the number -6.85 in Figure 9-1(a) is the point located 6.85 units to the left of zero. The graph of the number 3 is the point located 3 units to the right of zero. On this number line we have used one unit to represent a difference of 1, and we could just as well have used one in., or one mm, or 1/8 in., or whatever was convenient.

The number line of Figure 9-1(a) could be used to represent physical quantities. For example, the points on the line might be used to represent temperatures or pressures. The line can be made horizontal or vertical. If the number line is made vertical, as shown in Figure 9-1(b), positive numbers increase in the upward direction. Negative numbers increase in absolute value in the downward direction. In either case, a point on the number line is a graph of a number. The point represents a number.

Experimental data usually consist of pairs of numbers in a table, each number of the pair being a measurement of one of the two physical quantities involved in the experiment. If we want to display the relationship between these two variables, we must find a way of representing two numbers. Thus we need two number lines. These lines are placed so that they cross each other at the zero of each one. This point of crossing is called the *origin*. The lines are made perpendicular to each other and are called *coordinate axes*. As shown in Figure 9-1(c), we designate one axis as the *vertical axis*, and the other as the *horizontal axis*.

To graph a point with these coordinate axes, we need two numbers. From a table of experimental data, we graph each pair of numbers as a point. Let us plot the point representing the first pair of numbers in Table 9-1.

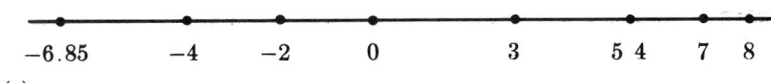

(a)
−6.85   −4   −2   0   3   5 4   7   8

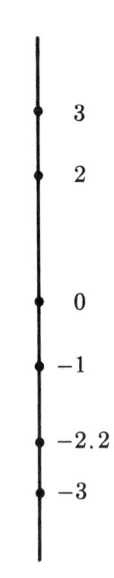

(b)
3
2
0
−1
−2.2
−3

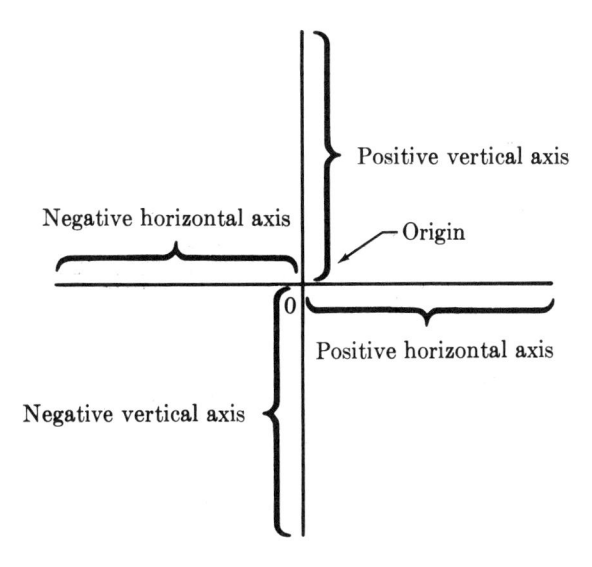

(c)

Figure 9-1.   (a) *Horizontal number line.*
(b) *Vertical number line.*
(c) *Coordinate axes.*

TABLE 9-1

V (volts)	I (amperes)
4.0	2.0
6.0	3.0
12.0	6.0
9.0	4.5

Voltage V is the dependent variable and current I is the independent variable. Usually, the dependent variable is placed on the vertical axis and the independent variable on the horizontal axis.

Before graphing the point representing the pair of numbers $(I,V) = (2.0, 4.0)$, we must prepare coordinate axes with proper labels. Figure 9-2 shows coordinate axes marked and labeled with the dependent variable V on the vertical axis and the independent variable I on the horizontal axis. Notice that we have shown what quantity is on each axis and in what units each quantity was measured. Graphs of measurements should always contain this information.

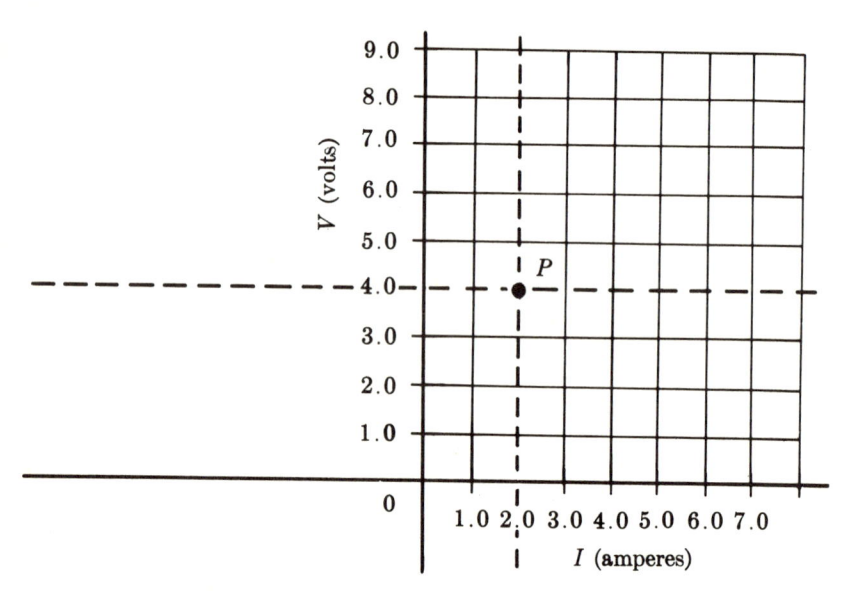

Figure 9-2. *The coordinates of a point.*

To graph the point representing the pair of measurements (2.0, 4.0), we first locate 2.0 on the horizontal axis. This first number is called the *abscissa* or *horizontal coordinate* of the point we are graphing. Moving to 2.0 on the horizontal axis, we imagine a vertical line passing through this point, extending parallel to the vertical axis indefinitely in both directions. Figure 9-2 shows that imaginary line with dashes. The current I is equal to 2.0 *everywhere* on this line; therefore, there are an infinite number of points along that line representing $I = 2.0$.

The other measurement, called the *ordinate* or *vertical coordinate*, is located along the vertical axis. Moving to 4.0 on the vertical axis, we imagine a horizontal line passing through this point, extending indefinitely in both directions. This line is also shown with dashes. Everywhere on this line, the voltage V is equal to 4.0, and again there are an infinite number of points having this value of V.

There are an infinite number of points on each of the two imaginary lines determined by the pair of measurements, but these lines intersect at only one point. That point of intersection is the only place where both $V = 4.0$ and $I = 2.0$. This point therefore represents the pair of numbers (2.0, 4.0). The point P in Figure 9-2 is therefore the **graph** of the first pair of measurements in Table 9-1.

We have used parentheses to describe the coordinates of a point. This practice is often followed. The first number in the parentheses is the abscissa, and the second number is the ordinate. As mentioned previously, the abscissa is the horizontal coordinate, and the ordinate is the vertical coordinate. To plot the graph of a point (I,V), we find the intersection of two imaginary lines, one vertical line passing through the horizontal axis at the value of I, the other a horizontal line passing through the vertical axis at the value of V.

PROBLEMS

1. Using a horizontal axis, locate each of the following measurements as points on the number line. Label the line carefully.
   a. $T = 8°$ b. $-3.7°$ c. $40°$ d. $-12°$

2. Construct a set of coordinate axes, labeling the vertical axis P (N/m^2) and the horizontal axis V (m^3). Graph each of the points formed by the following pairs of measurements.

P (N/m^2)	V (m^3)
5.3	1.80
2.5	3.82
1.1	8.66

3. Graph each of the following points.
   a. (2, 4)   b. (−12, 5)   c. (−5, −7)   d. (4, 0)   e. (8, −3).
   Which of these numbers are abscissas?
   Which are ordinates?

9-2 Graphs of Points; the Smooth Curve

We have graphed points representing pairs of numbers on coordinate axes. When an experiment provides a table of ordered pairs of measurements, we assume that if measurements were made in-between the ones we have, they would follow the same relationship. This assumption means that our limited number of pairs can be extended to an infinite set. The graph of isolated points is made into a smooth curve passing through each of the points of actual measurements. Figure 9-3 illustrates the correct and incorrect ways of graphing these points.

There are always uncertainties in measurements. Because of uncertainties, graphs of points cannot be merely points. Each point must have a width corresponding to twice the absolute uncertainty of the horizontal variable, that is $1\delta X$ more than the value and $1\delta X$ less than the value. The point must also have a vertical thickness of twice the absolute uncertainty in that value. Figure 9-4 shows a method of indicating these uncertainties. The pair of measurements represented by the point is $V = (5.2 \pm 0.3)$ v and $I = (7.9 \pm 0.4)$ amp.

Figure 9-4(b) shows graphs of the points for the measurements in Table 9-2. Notice that if we just showed the points, without indicating uncertainties, the smooth curve would be that shown by the dashed line. But when uncertainties are considered, a straight line curve passes through each rectangle. Our uncertainties tell us that we do not know where in the rectangle the actual point lies; it could be anywhere in it. The simplest "curve" is a straight line. Since a straight line passes through every rectangle, we assume that the relationship between V and I is *linear*, a straight line relationship. If our uncertainties had been small enough that the rectangles would not contain a straight line, then the relationship might have been a more complicated curve. This example illustrates the importance of making careful estimates of uncertainties.

TABLE 9-2

V (volts)	I (amperes)
0.5 ± 0.3	0.8 ± 0.2
0.7 ± 0.3	2.0 ± 0.2
1.6 ± 0.3	2.8 ± 0.2
1.9 ± 0.3	4.0 ± 0.2
2.6 ± 0.3	5.3 ± 0.2

(a)

(b)

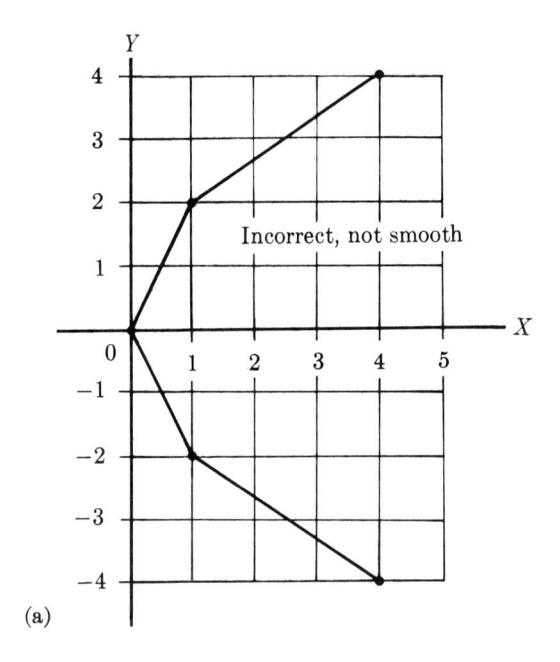

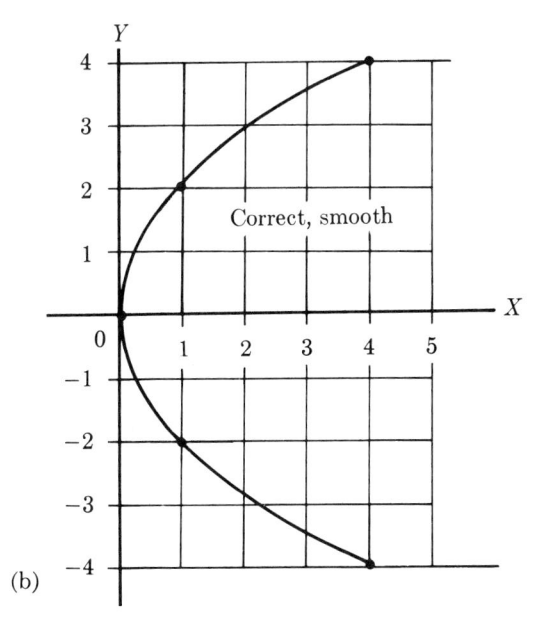

Figure 9-3. (a) *Incorrect, not smooth.*
(b) *Correct, smooth.*

(a)

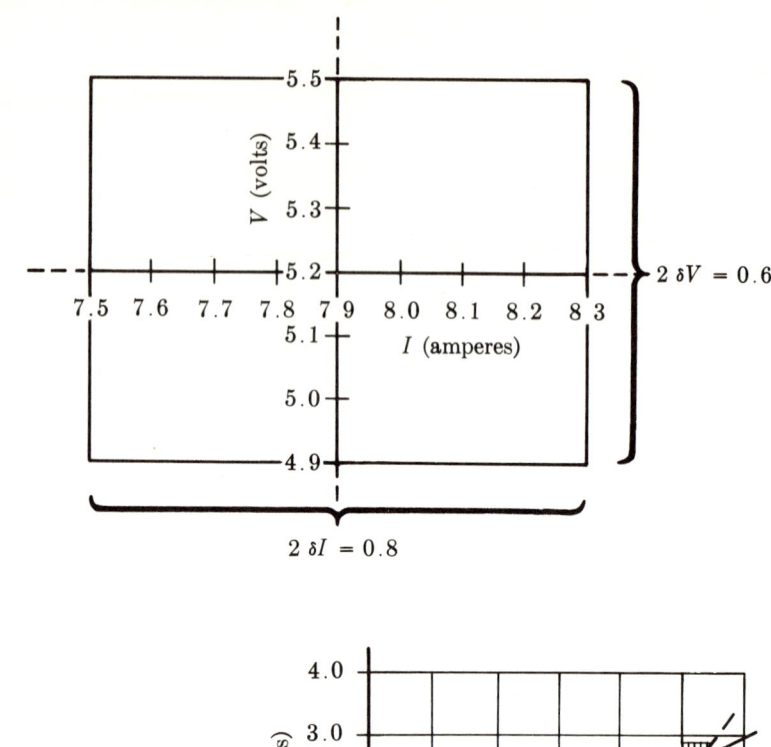

(b)

Figure 9-4. *Uncertainties determine sizes of coordinates.*

In most of the various graphs discussed so far, the pairs of numbers have been positive, but we have drawn our coordinate axes symmetrically; we have included the origin in each of the graphs shown. In practical situations, however, where graphs quite often involve only positive quantities, we need only show one quadrant of the coordinate system. Also, if we have experimental variables made over a small range, but with large values, we need not include the origin in our graph. As an example, consider the data in Table 9-3. If we plot the points representing these measurements on a complete coordinate system as shown in Figure 9-5, we have points much too close together. Yet, if we consider only a part of the coordinate system over the range of our measurements as shown in Figure 9-6, the graph has a more reasonable size.

TABLE 9-3

X (meters)	F (newtons)
0.48205	3.110
0.48360	3.120
0.48515	3.130
0.48670	3.140

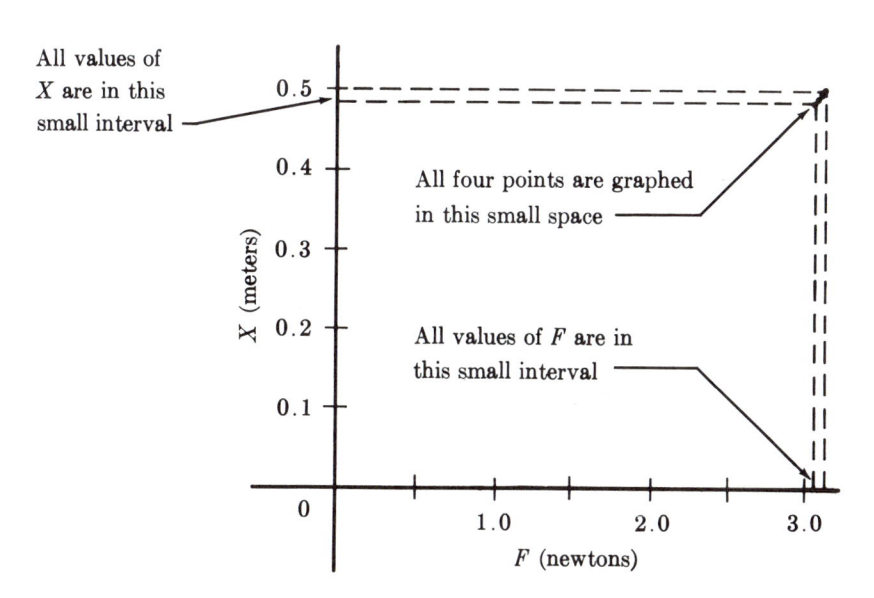

Figure 9-5. *The scale is not correct.*

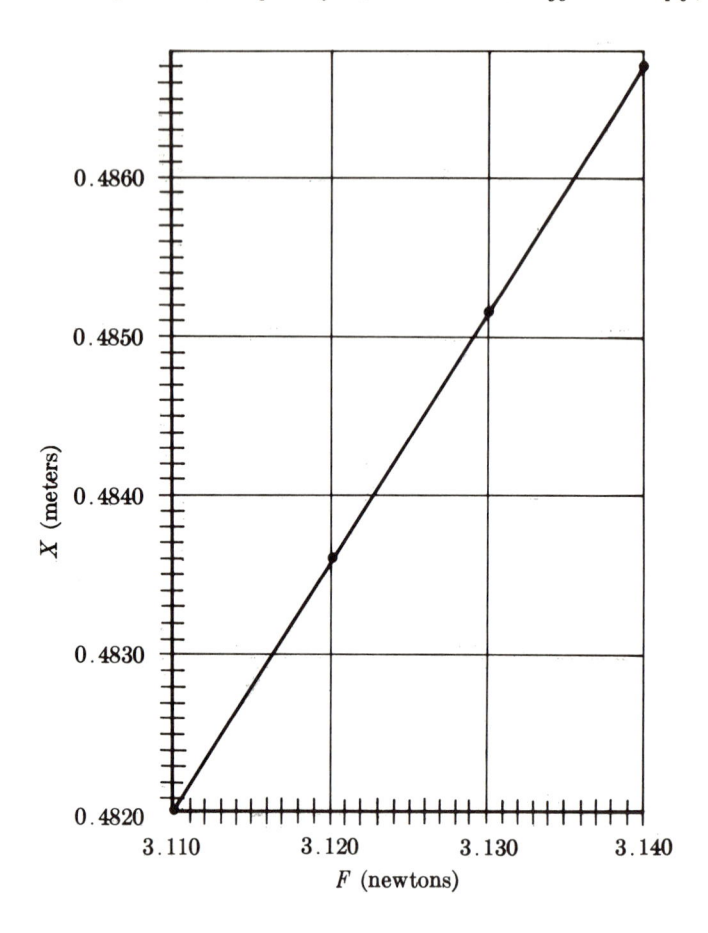

Figure 9-6. *An expanded scale.*

Before plotting a set of data from a table, the range of each variable should be determined, so that the scale dimensions on the graph paper can be marked to take in most of the sheet. Absolute uncertainties in the abscissas and ordinates should be used to construct rectangles centered on each point. The simplest curve is a straight line, and if a straight line will pass through all rectangles, this curve should be used. Otherwise, the best smooth curve should be constructed. The curve represents the relationship between the experimental variables for which the data were taken.

PROBLEMS

1. On a full sheet of graph paper, graph the points for the following data. Use the uncertainties to construct rectangles about each point; then draw the best smooth curve through the rectangles. Decide on the scale dimensions which will allow your curve to cover most of a single sheet of graph paper.

X (meters)	F (newtons)
-4.0 ± 0.2	-3.1 ± 0.1
-2.4 ± 0.2	-1.7 ± 0.1
-0.7 ± 0.2	-0.6 ± 0.1
0.6 ± 0.2	0.6 ± 0.1
4.3 ± 0.2	3.1 ± 0.1
7.1 ± 0.2	5.3 ± 0.1
12.7 ± 0.2	9.1 ± 0.1

2. Graph the points for the following data on a full sheet of graph paper, constructing rectangles using uncertainties, selecting appropriate scale dimensions, and drawing the best smooth curve through the points. Is the curve a straight line?

$P\,(N/m^2)$	$V\,(m^3)$
500 ± 5	43.0 ± 0.2
540 ± 5	39.7 ± 0.2
580 ± 5	37.0 ± 0.2
640 ± 5	33.5 ± 0.2
678 ± 5	31.7 ± 0.2
715 ± 5	30.0 ± 0.2

9-3 *Graphs of Linear Functions*

Many relationships between two experimental variables are of the type which can be represented by a *linear function*. A linear function has on each side of its defining equation only one first power variable. Algebraically, the linear function would be of the type $Y = KX + C$, where Y and X are variables, and K and C are constants.

We will construct a graph of a specific linear function; then we can look at a graph of the general linear function. For a specific example, let $K = 3$ and $C = -5$. Then we have the equation $Y = 3X - 5$. To graph this function, we must have a table of ordered pairs of numbers for points. To make this table, we substitute numbers for X and evaluate the expression for Y. Let us select for X the numbers 0, 1, 2, 3, and 4. To find values for Y, we substitute each of these values of X into the equation $Y = 3X - 5$. For example, using $X = 2$, we get

$$Y = (3)(2) - 5 = 6 - 5 = 1$$

The resulting pairs of numbers are shown in Table 9-4.

TABLE 9-4

Y	X
−5	0
−2	1
1	2
4	3
7	4

Graphing points of the numbers in Table 9-4 and drawing a smooth curve, we see in Figure 9-7 that this linear function is a straight line. Actually, all linear functions graph as straight lines. This is one of their properties. Before we examine certain features of this graph, however, we need to make two important definitions.

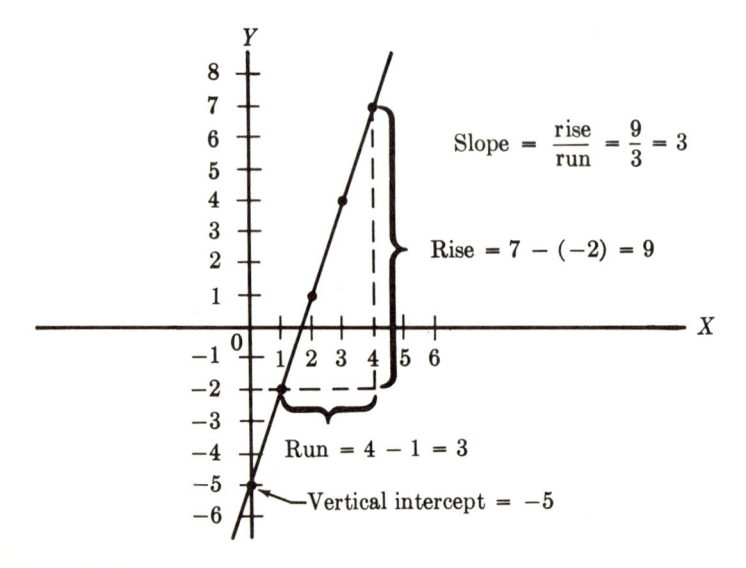

Figure 9-7. *The linear function: a specific example.*

[§ 9-3]   *Graphs of Linear Functions*   **171**

The first of these definitions is called *slope*. In simple terms the slope of a line is how far the line rises divided by the "run" corresponding to that rise. If a line is horizontal, its rise is zero, thus its slope is zero. If we imagine walking from left to right along the line, and we are walking uphill, the slope is positive. If we are walking downhill, the slope is negative. We could define slope as rise/run, but we should define these words in terms of symbols. To compute the slope, we select two points on the line, P_1 and P_2. The rise is found by taking $V_2 - V_1$ and the run by $H_2 - H_1$, where V_2 is the vertical coordinate of P_2, V_1 is the vertical coordinate of P_1, H_2 is the horizontal coordinate of P_2, and H_1 is the horizontal coordinate of P_1. Our definition for slope then becomes

$$\text{slope} = (V_2 - V_1)/(H_2 - H_1)$$

These differences, $V_2 - V_1$ and $H_2 - H_1$, are usually symbolized by ΔV and ΔH, respectively. This symbol Δ is another Greek letter, delta. We now have a definition for the slope, K, of a line:

$$K = (V_2 - V_1)/(H_2 - H_1) = \frac{\Delta V}{\Delta H}$$

In finding these differences, it must be remembered that a negative number like -2 is *larger* than a negative number like -12, and that -2 is *smaller* than 0. A smaller number is to the left of another number on the horizontal axis, or below another number on the vertical axis. Using this definition for the straight line graph of Figure 9-7, we get a slope of $K = 3$. This number is a measure of the steepness of that line. It would not make any difference which two points on the line were selected to determine the slope; the results would be identical. The slope is the same everywhere on a straight line.

The other definition which is frequently used for graphs of linear functions is called the *vertical intercept*. This quantity is just what the words imply; the vertical intercept is the value of the vertical coordinate where the line intercepts the vertical axis. Of course the horizontal coordinate is everywhere zero on the vertical axis; therefore, we could define the vertical intercept as the value of the vertical coordinate when the horizontal coordinate is zero. Figure 9-7 has a vertical intercept at $Y = -5$, thus -5 is the vertical intercept.

Returning to the general linear function $Y = KX + C$, we shall graph this equation. Because we know the equation is a straight line, we need to find only two points for our table. The easy way to find these points is to let $X = 0$, from which we get

$$Y = (K)(0) + C = C$$

and let $Y = 0$, from which we get

$$0 = KX + C$$

Solving this equation for X, we get

$$X = -C/K$$

Thus we have the coordinates $(0, C)$ and $(-C/K, 0)$ for our graph. Figure 9-8 shows that graph.

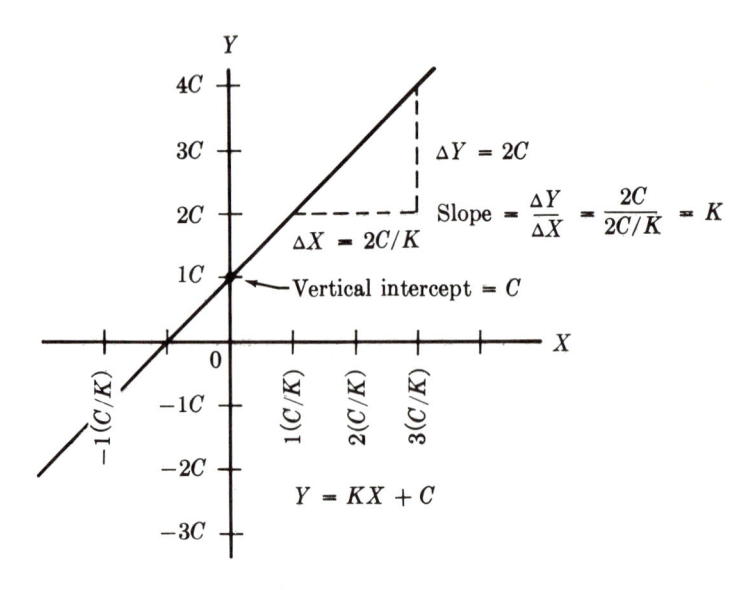

Figure 9-8. *The general linear function.*

Finding the slope of the graph of Figure 9-8, we get slope = K. We see also that the vertical intercept is equal to C. These two results for the general linear equation are very important. If you are given some equation like $Y = 9X + 3$, without even graphing it you know that the slope is 9 and the vertical intercept is 3. Perhaps more important in experimental work, if a table of data graphs as a straight line, you can make an equation from the graph just by finding the slope and the vertical intercept. Although we plotted the graph in Figure 9-7 by making a table from the equation $Y = 3X - 5$, which we might write as $Y = 3X + (-5)$, the equation tells us that the slope should be 3 and the vertical intercept -5. These are the values we actually found in Figure 9-7.

(a)

(b)

(c)

(d)

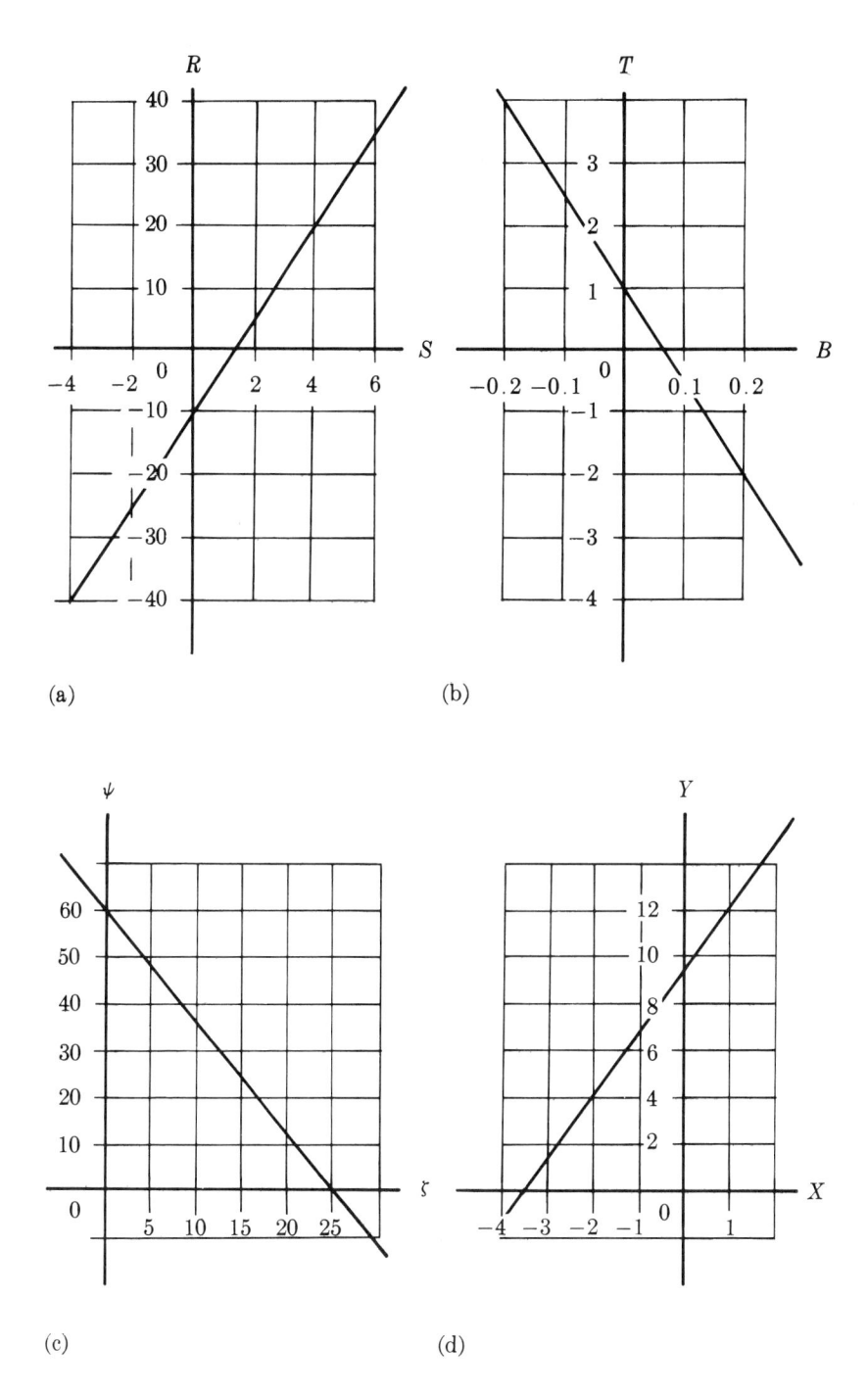

Figure 9-9.   *Graphs of specific linear functions.*

173

A linear function graphs as a straight line. The general equation for the linear function in the slope-intercept form is

$$Y = KX + C$$

Defining slope as rise divided by run, or symbolically as $\Delta V/\Delta H$, where $\Delta V = V_2 - V_1$ and $\Delta H = H_2 - H_1$, we found that K in the equation is the slope of the graph. Defining the vertical intercept as the value of the vertical coordinate where the line crosses the vertical axis, we found that the vertical intercept was equal to C in the equation. These two results enable us to graph linear equations quickly, or to construct equations from graphs of experimental data.

PROBLEMS

1. Using the equation $Y = -4X + 5$, substitute values for X and make a table containing at least 5 pairs of numbers. Graph these points on a full sheet of graph paper, labeling the axes and each point. Label the points with horizontal coordinate and vertical coordinate in parentheses next to the point.

2. Find the slope of the graph of Figure 9-6.

3. Find the slope and vertical intercept of the graph of Figure 9-4 (b), making an equation relating the variables V and I.

4. For the table of data in Problem 1 of Section 9-2, find the slope of the graph and the vertical intercept. Construct an equation relating the variables X and F.

5. What is the slope and vertical intercept of graphs for each of the following equations?
   a. $Y = -5X - 6$ b. $Y = 4X + 4$ c. $Y = 2X - 9$
   d. $Y = -8X + 7$ e. $4X - 3 = 5Y$ f. $8Y - 6 = 22X$

6. For each of the graphs of Figure 9-9, use the slope and vertical intercept to find an equation of the line.

9-4 Linear Plots of Non-linear Functions

The preceding section provided a powerful means for analyzing experimental data for which graphs were linear. But many simple experimental relationships are non-linear. Figure 9-3 is a graph of a non-linear function; it is not a straight line. Fortunately, there is a method by which some non-linear functions can be made linear.

[§ 9-4]   Linear Plots of Non-linear Functions   **175**

The data for Figure 9-3 is shown in Table 9-5. To modify the data to make it plot linearly, we must first guess the relationship. It is a trial and error procedure. Because of the symmetry of the graph of Figure 9-3(b) about the X-axis, Y could equal some number times the square root of X. Stated another way, we might have a non-linear relationship like $Y^2 = KX$. If this relationship is correct, we should be able to square each value of Y in the table; then when Y^2 is plotted against X, we should get a straight line with the slope equal to K.

TABLE 9-5

Y^2	Y	X
16	4	4
4	2	1
0	0	0
4	−2	1
16	−4	4

Figure 9-10 shows the graph of Y^2 on the vertical axis and X on the horizontal axis, using the data of Table 9-5. The graph is a straight line, the slope being equal to 4. Thus we know that the relationship is $Y^2 = 4X$.

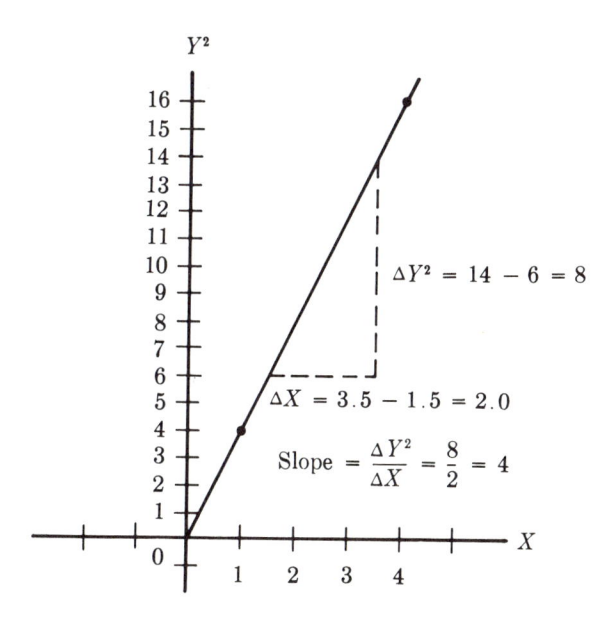

$\Delta Y^2 = 14 - 6 = 8$

$\Delta X = 3.5 - 1.5 = 2.0$

$\text{Slope} = \dfrac{\Delta Y^2}{\Delta X} = \dfrac{8}{2} = 4$

Figure 9-10.   *A non-linear function made linear.*

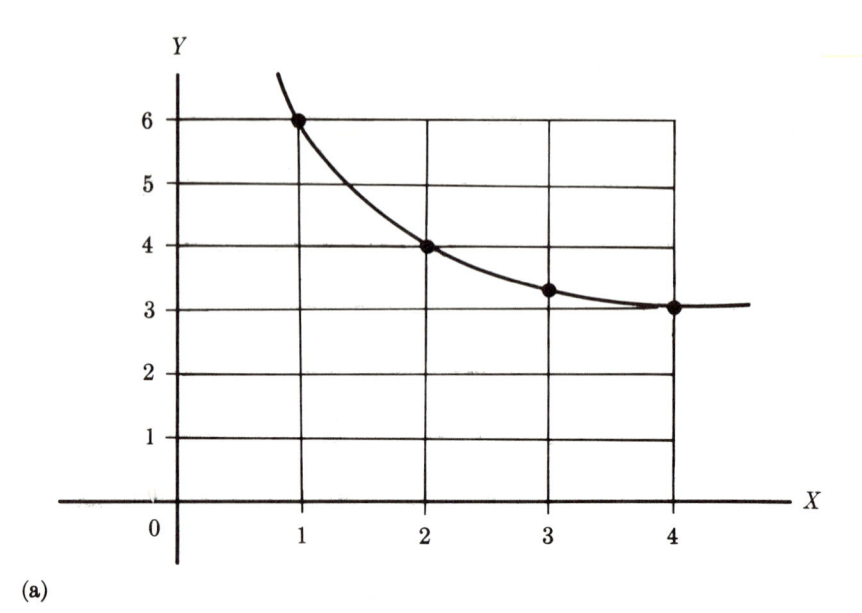

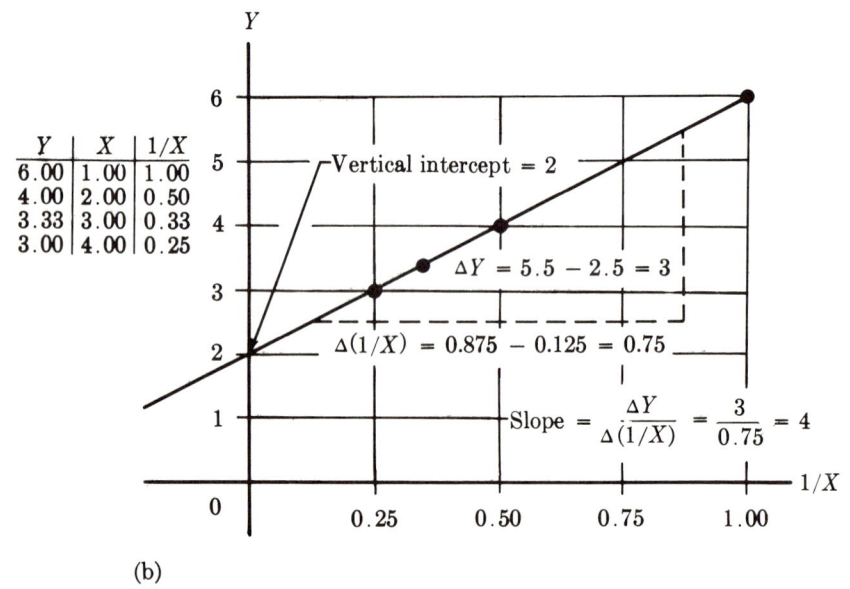

Figure 9-11. *A non-linear function made linear.*

To make a non-linear relationship plot as a linear function, we must guess the correct form of the equation. It is rather like having to know the answer to find the answer. But the graph serves to confirm your guess. If you have not guessed correctly, you will not get a straight line. Fortunately, the process is not entirely guesswork. If one variable increases as the other decreases, you know that the relationship is in some way inverse. If they increase together, you know it is some kind of direct relationship. If a graph appears to be parabolic, one of the variables could be squared. If the curve appears to be a hyperbola, the relationship might be a simple inverse proportion. Experience with such graphs teaches one to make very good guesses.

As a final example, let us examine the data in Table 9-6. We notice that as X increases, moving down the table, Y decreases. From this brief analysis, we suspect some kind of inverse relationship. The graph of these data in Figure 9-11 (b) shows a plot of Y on the ordinate and $1/X$ on the abscissa. Notice that we do get a straight line, with a slope of 4 and a vertical intercept of 2. Thus the relationship must be $Y = 4(1/X) + 2$, or, more simply, $Y = 4/X + 2$.

TABLE 9-6

Y	X
6.00	1.00
4.00	2.00
3.33	3.00
3.00	4.00

The method of making linear plots of non-linear functions can be useful when a certain relationship is suspected, or when relationships are somewhat complex. For some kinds of functions, however, a quick plot on a special kind of graph paper can quickly reveal details of the relationship. We shall examine these techniques in the following sections.

PROBLEMS

1. Using the data of Problem 2 at the end of Section 9-2, plot a graph of P on the vertical axis and $1/V$ on the horizontal axis. What is the equation which gives the relationship between P and V?

2. An experiment is performed to determine the relationship between the period T of a pendulum and the length L of the pendulum. For the following data, plot a graph of T on the vertical axis and L on the horizontal axis. Is the relationship linear? If the result is non-linear,

plot other graphs of T on the vertical axis, and quantities like L^2, $1/L$, $\sqrt{L}$, etc., on the horizontal axis, until you find a linear relationship. Write the equation of this relationship.

T (seconds)	L (meters)
2.01 ± 0.01	1.000 ± 0.002
1.79 ± 0.01	0.800 ± 0.002
1.56 ± 0.01	0.600 ± 0.002
1.27 ± 0.01	0.400 ± 0.002
0.90 ± 0.01	0.200 ± 0.002

3. For the data in the following table, try different plots of Y on the vertical axis, and X, X^2, $1/X$, $1/X^2$, etc., on the horizontal axis until you find a linear relationship between Y and one of these functions. From the slope and vertical intercept of your linear function, write the equation relating Y and X.

Y	X
8	0
3	1
-12	2
-37	3

4. A research scientist is studying the relationship between the inverse spin-lattice relaxation time, τ^{-1}, and the absolute temperature, T, for certain crystals at extremely low temperatures. He suspects a relationship of the form $\tau^{-1} = AT + BT^7$. He would like to determine whether this is the correct equation and, if it is, what the values of A and B are. To get an equation in a form that would plot linearly, he can divide both sides of this equation by T, getting $\tau^{-1}/T = A + BT^6$. Thus, if he divides each value of τ^{-1} by the corresponding value of temperature T, and also finds T^6, a graph of τ^{-1}/T on the vertical axis and T^6 on the horizontal axis should give a straight line with a vertical intercept of A and a slope of B. Use data in the following table to plot that graph. What are the values of A and B? What is the equation for τ^{-1}?

τ^{-1}	T	τ^{-1}/T	T^6
1452	1.11	1308 ± 5	1.870 ± 0.003
1511	1.12	1350 ± 5	1.975 ± 0.003
1563	1.13	1384 ± 5	2.084 ± 0.003
1642	1.14	1440 ± 5	2.200 ± 0.003

9-5 Semi-logarithmic Graph Plots

The various techniques of graphical analysis discussed so far in this chapter have been concerned with linear functions or non-linear functions made linear by appropriate guesses. The graphs have been made on ordinary linear graph paper. But there is some advantage in using a special kind of graph paper which eliminates some of the guess-work in our search for certain kinds of relationships. This non-linear graph paper automatically gives straight lines. Two of the types of graph paper that have non-linear axis spacing are *semi-logarithmic* and *logarithmic-logarithmic*. To use such graph paper, an understanding of logarithms is essential.

Of all the topics in elementary mathematics, logarithms are probably the least understood. Yet considering the essential simplicity of the concept of logarithms, any confusion about them is sadly unnecessary.

In elementary algebra, the student learns a few simple rules for exponents. The basic rules are these:

$$X^n X^m = X^{n+m} \qquad (9\text{-}1)$$

$$X^n / X^m = X^{n-m} \qquad (9\text{-}2)$$

$$(X^n)^K = X^{nK} \qquad (9\text{-}3)$$

Few students would have even momentary difficulty understanding or using these rules for exponents, and there is nothing in principle about logarithms that involves more than these three simple rules. One only needs to understand that *a logarithm is an exponent*. Perhaps the confusion would never have been so severe if the word *exponent* were used instead of the word *logarithm*.

If a logarithm is an exponent, then when two logarithms are added, we must be multiplying two numbers. If a difference of two logarithms is being found, then we are dividing two numbers. If a logarithm is multiplied by K, then we are raising some number to a power K. These facts about logarithms are consequences of the three simple rules for exponents.

We define logarithm in the following way. Using the abbreviation *log* for the word logarithm, the statement $Y = \log_a X$ means *the exponent needed on base a to give the number X is Y*. Everytime you see logarithms, you should *think* about them using the words above. As a simple example, changing $4^2 = 16$ to logarithmic form, we have $\log_4 16 = 2$. Notice how this equation is read: the exponent needed on a base of 4 to give 16 is 2. You can see that using logarithms merely states an equation with exponents in a different way.

With an understanding that logarithms are exponents, we can state the three simple rules for exponents as rules for logarithms. Those rules are:

$$\log_a X + \log_a Y = \log_a(XY) \qquad (9\text{-}4)$$

$$\log_a X - \log_a Y = \log_a(X/Y) \qquad (9\text{-}5)$$

$$\log_a X^m = m \log_a X \qquad (9\text{-}6)$$

Equation (9-4) follows directly from the algebraic rule for multiplying numbers that have the same base by adding exponents. The second rule, Equation (9-5), is correct because when two numbers are divided, exponents are subtracted. The third rule can be proved by writing $Y = \log_a X^m$ in exponential form, that is, as $a^Y = X^m$. Then, taking both sides of this equation to the $1/m$ power, we have

$$a^{Y/m} = X$$

Rewriting this equation in logarithmic form gives

$$Y/m = \log_a X$$

Finally, multiplying both sides of this equation by m, we get

$$Y = m(\log_a X) = \log_a X^m$$

Semi-logarithmic graph paper has one coordinate axis in exponental form, the other axis being linear. When a number is located on the log axis, the logarithm of that coordinate is proportional to the distance from the other axis to the coordinate. In effect, you are automatically taking the logarithm of that coordinate when you locate the point.

To understand how semi-logarithmic graph paper can be useful, consider an equation of the form

$$Y = Ka^X$$

Dividing both sides of this equation by K, and expressing the result in logarithmic form, we have

$$\log_a(Y/K) = X$$

Using Equation (9-5), we have

$$\log_a Y - \log_a K = X$$

or

$$(\log_a Y) = X + (\log_a K)$$

This equation shows that if a relationship of the type $Y = Ka^X$ were plotted on semi-log paper of base a, the graph would be a straight line, with the slope equal to 1 and a vertical intercept at $\log_a K$. Since the graph paper automatically takes logarithms, the value of K can be read directly from the graph.

Semi-logarithmic paper is usually constructed using a base of ten. For that reason, when we express the equation $Y = Ka^X$ in logarithmic form, we get

$$(\log_{10} Y) = X(\log_{10} a) + (\log_{10} K)$$

From this equation, we see that the vertical intercept still reads the value of K directly, but the slope is equal to $\log_{10} a$. Thus, to evaluate a we must use tables of logarithms to the base ten, looking up the exponent needed on ten to give the value a.

Let us examine a specific example. We shall start with the equation

$$Y = (3)(5^X)$$

Substituting values for X, we can compute values of Y to make Table 9-7.

TABLE 9-7

Y	X
15	1
75	2
375	3
1875	4

These pairs of numbers are plotted on semi-logarithmic graph paper in Figure 9-12. In plotting these points, we are automatically taking the logarithm of each Y value, but the X-coordinates are being plotted on an ordinary linear axis.

The graph of the data in Table 9-7 does form a straight line on semi-log paper. For such a straight line, we would expect the vertical intercept reading to be equal to K in the equation $Y = Ka^X$. The vertical intercept reading is the number 3, as expected. To find the slope of this graph, we must divide the rise by the run. But how do we find the rise? It is not found from a reading of the coordinates. We must measure the rise with a scale, dividing this measured result by a similar measurement of the base of the graph paper, that is, a measurement from one to ten on the vertical axis.

We find a slope of 0.699 using the procedure described above. Previously, we had shown that the slope was equal to $\log_{10} a$. Looking up 0.699 in a table of logarithms of base ten, we find that 0.699 is the exponent needed on ten to give 5. The quantity a in the equation $Y = Ka^X$ is therefore equal to 5. We have analyzed this semi-log graph and found that the curve is that of the equation

$$Y = (3)(5^X)$$

which is precisely the equation we used in constructing the graph in the first place.

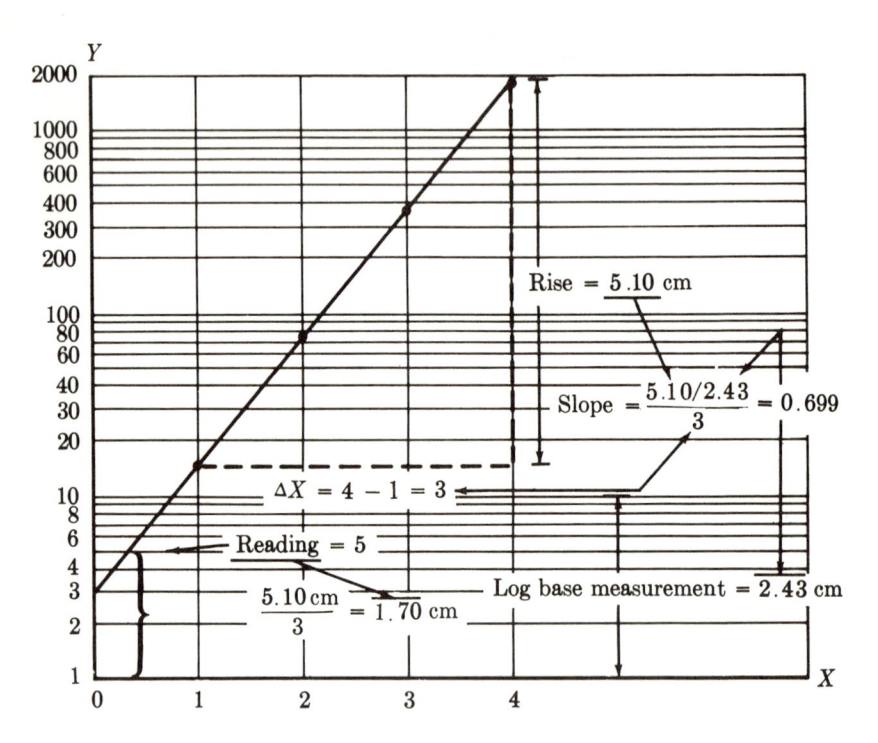

Figure 9-12. *A semi-logarithmic graph plot.*

It should be noted that semi-log graph paper can be used as a "table of logarithms" of base ten. This property of the paper enables the slope measurement to be made in the following way. Measure the rise in centimeters, inches, or in whatever units are convenient. Divide this measurement by the run, found in the conventional way using $H_2 - H_1 = \Delta H$. Now, instead of dividing this result by the log base measurement, just lay off this distance along the vertical log axis, starting at the origin. When this is done, the reading on the vertical axis is the value of a in the equation $Y = Ka^x$. This method is illustrated in the lower left hand region of Figure 9-12.

In the example just worked, we knew the equation before constructing the graph. Of course, if we knew the equation, there would be no need to make a graph to determine it. But in a real situation we have a table of data, but we have no idea what the equation might be. Let us

[§ 9-5] Semi-logarithmic Graph Plots 183

use the same technique developed above to find the equation relating the data in Table 9-8.

TABLE 9-8

Y	X
120	1
720	2
4320	3

The three points of Table 9-8 are graphed in Figure 9-13. As shown in the figure, the vertical intercept gives the value of K, and the slope the value of a, in the equation $Y = Ka^X$. Thus, from our graphical analysis, we get

$$Y = (20)(6^X)$$

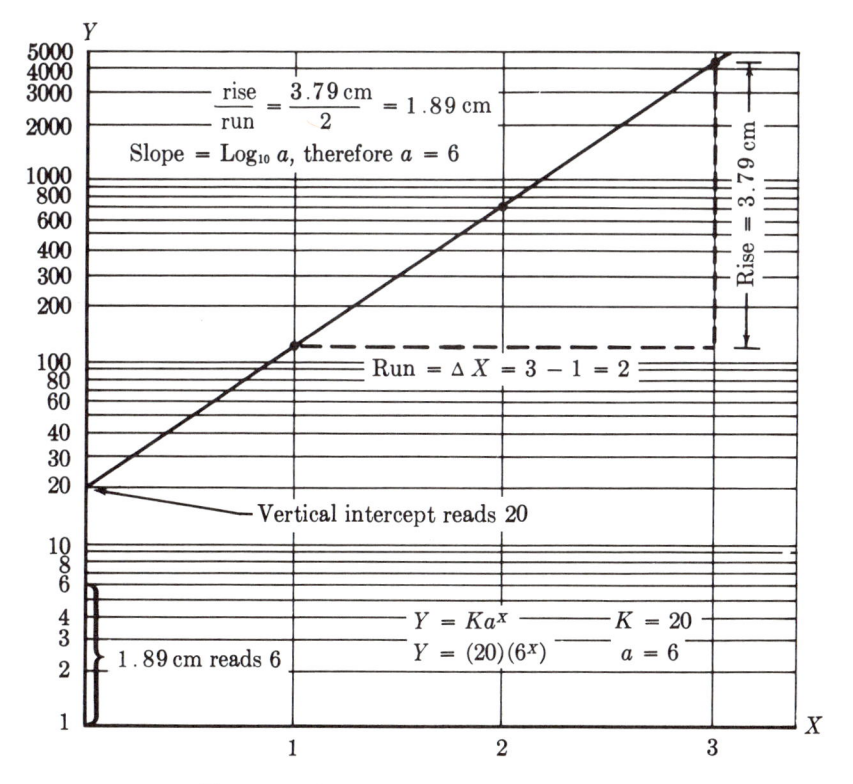

Figure 9-13. *Reading a semi-log graph.*

Besides using semi-log graph paper for analyses like these, such paper can be used to get a graph of some relationship where one of the coordinates covers a very large range, from the very small to very large.

PROBLEMS

1. Change each of the following equations to logarithmic form.
   a. $Y = 6^X$ b. $X = a^Y$ c. $25 = 5^2$ d. $0.25 = 2^{-2}$
   e. $100{,}000 = 10^5$ f. $8 = 10^{0.9031}$

2. Change each of the following equations to exponential form.
   a. $Y = \log_b X$ b. $-3 = \log_2 0.125$ c. $4 = \log_5 625$
   d. $X = \log_b Y$ e. $0 = \log_{12} 1$ f. $0 = \log_X 1$

3. Evaluate each of the following logarithms.
   a. $\log_4 64$ b. $\log_2 64$ c. $\log_{10} 10{,}000$ d. $\log_{12} 12$
   e. $\log_5 3125$ f. $\log_{10} 0.00001$

4. A graph of a set of data plots as a straight line on semi-logarithmic graph paper. The vertical intercept reads 7. The rise is measured at 12.12 cm, the run being equal to 4. The distance 12.12 cm/4 = 3.03 cm measured from the origin on the vertical axis gives a reading of 3. What equation relates the data of this experiment, if Y is on the vertical (log) axis and X is on the horizontal (linear) axis?

5. A hi-fi enthusiast wants to determine how good his output transformer is in providing all audio frequencies to his speaker system at the same level. He feeds the input of the transformer with a variable source of audio frequencies from an oscillator that produces signals which are extremely constant in sound level for all frequencies. When he measures the output of his transformer, he gets the following data.

f(cps)	V (volts)
100	8.0
300	9.0
500	10.0
700	10.0
1000	10.0
3000	10.0
7000	10.0
12000	10.0
18000	9.6
20000	8.3
25000	5.2

Construct a graph of these data, plotting frequencies on the log axis of semi-log paper, with output voltages on the linear axis. How "flat" is his transformer? Why would ordinary graph paper not have been appropriate for this kind of data?

6. A radioactive source decays (sends out radiations) at a rate given by the equation

$$R = R_0 e^{-\lambda t}$$

where R is the decay rate at time t, R_0 was the decay rate at time zero, λ is a constant, and e is the base of the natural logarithms. The value of e is approximately 2.7183. The equation for the total number N of decays at time t is given by

$$N = N_0 e^{-\lambda t}$$

where N_0 is the original number of excited atoms. When half of the atoms have decayed, at some time T called the *half-life*, we must have

$$N_0/2 = N_0 e^{-\lambda T}$$

Dividing both sides of this equation by N_0, we have

$$1/2 = e^{-\lambda T}$$

Taking the logarithm to the base ten of both sides of this equation we have

$$\log_{10}(1/2) = \log_{10}(e^{-\lambda T})$$

Now using Equation (9-5) on the left side, and Equation (9-6) on the right side, we get

$$\log_{10}1 - \log_{10}2 = -\lambda T(\log_{10}e)$$

But $\log_{10}1 = 0$, $\log_{10}e = 0.43429$, and $\log_{10}2 = 0.3010$, thus

$$T = (0.3010)/(\lambda 0.43429) = 0.693/\lambda$$

The half-life T is related to λ by the equation

$$T = 0.693/\lambda$$

If we had some way of finding λ, we could compute the half-life of a radioactive source. But the original rate equation tells us that all we need to do is plot a graph of rate R on the log axis of semi-logarithmic graph paper, time being plotted on the linear axis. The slope of this graph would then give $\log_{10}e^{-\lambda}$, which by Equation (9-6) can be written $-\lambda \log_{10}e$. Since $\log_{10}e = 0.43429$, the slope of the graph is just $(-\lambda)(0.43429)$, and we can find λ. The following data is from an experiment with a radioactive source. Graph the data on semi-log paper and find λ. From λ, compute the half-life T.

R (counts/min)	t (min)
100	0
75	5
56	10
42	15
32	20

9-6 Log-Log Graph Plots

Some graph paper is available which has logarithmic spacing on both coordinate axes. This log-log graph paper automatically takes the logarithm of each coordinate. In this way, certain kinds of functions graph as straight lines, the slope and vertical intercept giving details of the equation of the relationship.

Consider an equation of the form $Y = KX^n$. If we take the logarithm to the base ten of both sides of this equation, we get

$$\log_{10} Y = \log_{10}(KX^n)$$

But, by Equation (9-4),

$$\log_{10}(KX^n) = \log_{10} K + \log_{10} X^n$$

Then, by Equation (9-6),

$$\log_{10} X^n = n(\log_{10} X)$$

Thus we have

$$(\log_{10} Y) = n(\log_{10} X) + (\log_{10} K)$$

In terms of logarithms, this equation is in the form

$$y = nx + b$$

therefore, n is the slope and K is read as the vertical intercept on log-log paper. The slope is dimensionless as long as the rise and run are measured in the *same* units.

As an example, we shall start with a known equation, $Y = 20X^2$. Substituting values for X, we solve for Y to get Table 9-9. These pairs of numbers are plotted on log-log graph paper in Figure 9-14, and they do form a straight line. The slope of this line is 2, and the vertical intercept is read as 20. Thus, as expected, the equation is $Y = 20X^2$.

TABLE 9-9

Y	X
80	2
180	3
320	4
500	5

[§ 9-6]   *Log-Log Graph Plots*   187

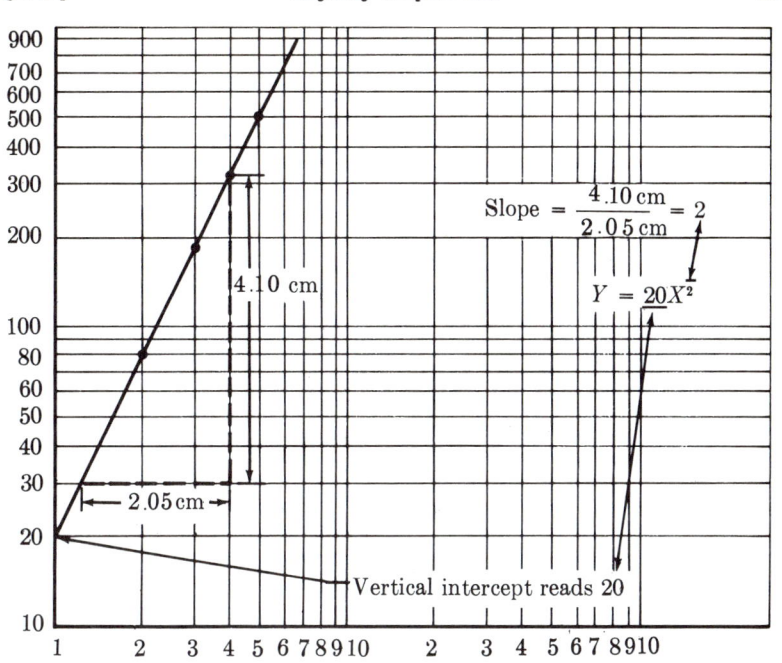

Figure 9-14.   *A log-log graph plot.*

As a more realistic problem, let us examine the data in Table 9-10. The table entry R (meters) represents the mean distance from the indicated planet to the sun. The entry T (earth days) is the period of revolution of the planet about the sun. The problem of analyzing these data was one that required some 25 years of the life of Johannes Kepler (1571-1630) before he discovered his third law of planetary motion.

TABLE 9-10

Planet	R (meters)	T (earth days)
Mercury	5.79×10^{10}	88.0
Venus	10.81×10^{10}	224.7
Earth	14.95×10^{10}	365.3
Mars	22.78×10^{10}	687.0

To plot these data on log-log paper, we need not be concerned about the 10^{10} factors, except to label properly our vertical axis. Figure 9-15 shows this graph. Notice that it is a straight line, with a slope of 2/3 and a vertical intercept at 1.37×10^{10}. Constructing an equation, we get

$$R = (1.37 \times 10^{10})T^{2/3}$$

Writing the equation another way by cubing both sides, we get

$$R^3 = (1.37 \times 10^{10})^3 T^2$$

and dividing both sides of this equation by T^2, we get

$$R^3/T^2 = (1.37 \times 10^{10})^3$$

which is constant for all planets in our solar system. This result, R^3/T^2 = constant, was Kepler's famous third law of planetary motion which took years for him to determine from data similar to those in Table 9-10.

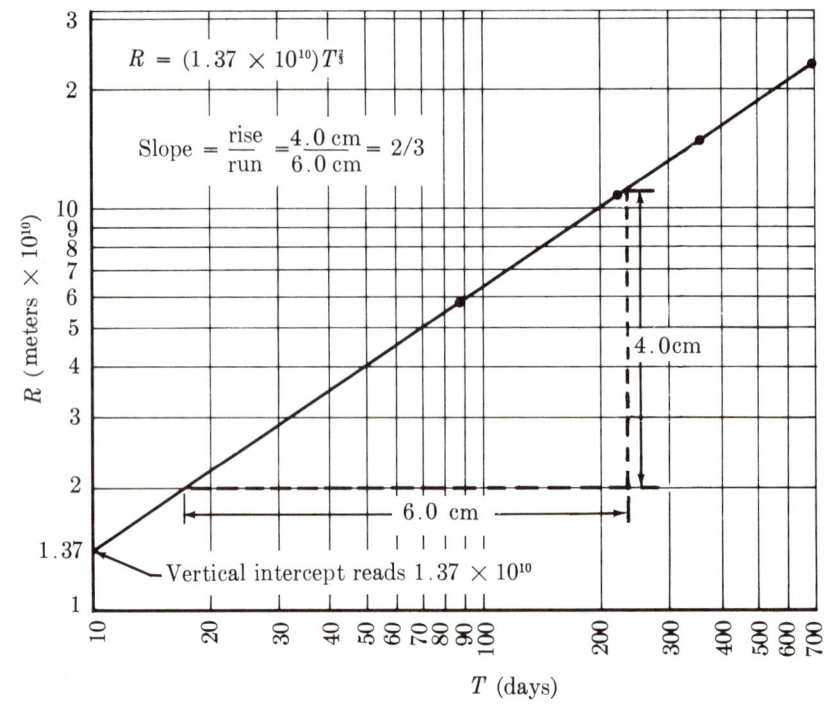

Figure 9-15. *Reading a log-log graph.*

The analysis of the data on planetary motion serves to illustrate the ease with which log-log paper can be used to find an otherwise elusive relationship between two variables. The basic technique is simple. Select log-log paper with sufficient cycles to cover the range of variables involved. Then plot points on the paper. Compute the slope from measurements of the rise and run made with any convenient instrument in any unit. This slope is the exponent in the equation $Y = KX^n$. The value of K is found by reading the vertical intercept.

PROBLEMS

1. Using the data in the table of Problem 2 at the end of Section 9-4, plot a graph on log-log paper and determine the equation relating T and L.

2. Using the data in Problem 2 at the end of Section 9-2, plot a graph on log-log paper and determine the equation relating P and V.

3. An experiment is performed in which electrical charges interact. The apparatus measures the force F between charges and the separation R of the charges. The following table of data was collected in that experiment. Construct a graph of these data on log-log paper; then find an equation relating R and F. What kind of law is this relationship?

F (millinewtons)	R (centimeters)
250.0	2.0
62.5	4.0
40.0	5.0
10.0	10.0

10 A Scientific Experiment

Many of the topics discussed in this book have necessarily been treated as somewhat isolated subjects, without their connection with each other being demonstrated. Now we shall examine a complete scientific experiment, for which the performance and analysis requires almost every quantitative ability you have learned so far. You must identify variables, make an experimental design, carry out a series of measurements, and analyze data, both numerically and graphically. Finally, you must summarize the results of the experiment.

A scientific experiment involves holding all but two variables constant; then, while one of these two is manipulated, changes in the other are observed. Before we can hold variables constant, or manipulate, or observe, we must select some problem for investigation.

Call Steve Vorhees
College Relations - 503.

↓↓↓

John needs count today from STD 101 as to who needs Math courses (121 & 124 particularly) next term.

JP

5.38

22.5×0.653
_____ = _____
   4.38

    268.5
_____ = _____
 4.96 × 1.723

 627 × 432
_____ = _____
 1.78 + 3.56

 927   625
_____ + _____ = _____
 432   221

The Area of an 8" DIAMETER CIRCLE = _____

7/523,468.52

$$\frac{5.38}{} =$$

$$\frac{22.5 \times 0.653}{4.38} =$$

$$\frac{268.5}{47.6 \times 1.723} =$$

$$\frac{627 \times 432}{1.78 + 3.56} =$$

$$\frac{927}{432} + \frac{625}{221} =$$

The Area of an 8" Diameter Circle =

$$7 \overline{) 523,468.52}$$

CLASS QUIZ STD 101

$5.35 \times 13.68 = \underline{73.1}$

$1.526 \times 279 = \underline{426,000}$

$(0.756)^2 = \underline{.573}$

$(12.56)^2 \times 13.6 = \underline{2140}$

$(0.075)^{.005} \times (0.152)^{.03} = \underline{.000130}$

$(17.23)^{297} + (22.9)^{525} = \underline{822}$

Long Hand

$$\begin{array}{r} 25.07 \\ \sqrt{628.51} \\ 4 \\ \hline 45 \overline{|228} \\ 225 \\ \hline 5007 \overline{|35100} \\ 35049 \end{array}$$

BY LONG HAND

One of the most common natural phenomena involves vibrating systems. The few laws responsible for oscillatory *simple harmonic motion* apply approximately to such diverse phenomena as the tones of all musical instruments, the pitch of a person's voice, and, in fact, to all sounds that we hear. Mechanical vibration is of considerable interest to the engineer, and the electrical analog involves the principle of tuned circuits, a major concept in electronics.

Before we can begin an investigation concerning vibrating systems, we must select some specific system that is amenable to study. If the variables that are common to all vibrating systems are identified, a law discovered for one system should apply to all.

10-1 Identification of Variables

To learn the basic laws of vibrating systems, we should select a system that is relatively simple, and for which we can control all variables. The characteristics common to all oscillatory systems involve a slight deformation or displacement from some equilibrium position. Then, when the system is released, vibration occurs which may be for a very short duration, or it may continue for some time. A guitar string plucked and released vibrates in this way. Other systems are continually displaced from equilibrium, but in such a way as to keep time with the vibration rate. This process keeps the system vibrating as long as the periodic displacements are imposed on it. Blowing air through ones lips to whistle involves this type of vibration.

A particularly simple vibrating system involves a mass suspended by a spring. When the mass is displaced and then released, it oscillates for quite a long time before coming to rest again. The system is easy to assemble, and measurements can be made without serious difficulty.

Now, what are the variables? To answer this question, some preliminary investigation is necessary. We notice that when a guitar string is tighter, the pitch is higher. When a spring is "stiffer", the vibration of a mass is faster. For the spring, then, some measure of stiffness must be one variable.

We also notice that for any stringed instrument, the pitch is lower for thicker strings, those having a larger mass per unit length. The larger the mass placed on the end of a spring, the slower the vibration rate. Mass or some measure of inertia is evidently a variable of importance.

Plucking a guitar string hard, and then easy, changes the loudness of the sound, but the pitch seems to remain the same. A mass on a spring appears to vibrate at about the same rate, whether it vibrates with a large amplitude, or barely moves. Yet, without actual measurements, we cannot be certain that amplitude is not a variable of importance. We certainly should investigate amplitude as a possible variable.

For all vibrating systems, the only variables which affect the period of vibration include some measure of the stiffness of the system, some measure of the inertia of the system, and perhaps the amplitude of vibration. We have identified these variables by making qualitative observations on vibrating systems.

For the specific vibrating system we shall examine in detail, the stiffness factor must involve the stiffness of the spring. The inertia measurement must be the mass attached to the end of the spring. Since part of the spring will also be vibrating, and the spring has mass, we must be sure in our experiment that the mass suspended is much larger than the mass of the spring. Then any errors introduced by ignoring the spring's mass will be of no consequence. The amplitude can be a measurement of how far the mass is displaced from its equilibrium position.

Figure 10-1 shows a spring being stretched an amount Y by some weight W. If the spring were very stiff, it would take more weight to stretch it a given distance than for a less stiff spring. We can therefore get a measure of the stiffness of the spring by finding out how much weight is needed to stretch the spring a unit distance.

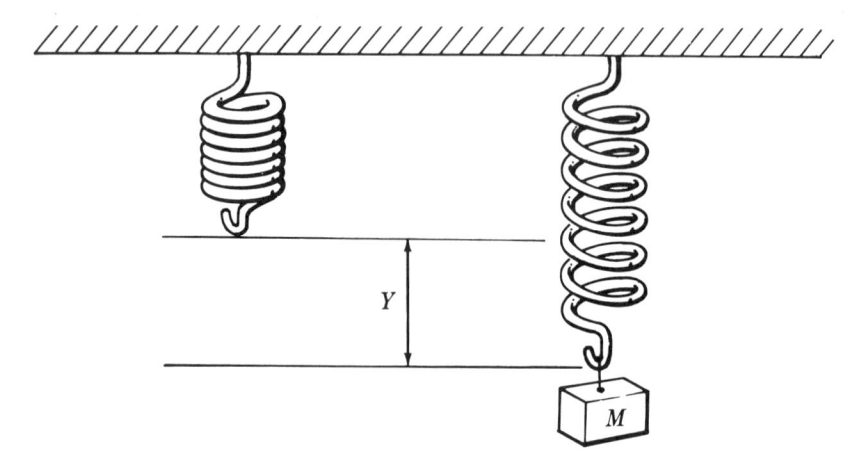

Figure 10-1. *A spring is stretched a distance Y by the weight of a mass M.*

If we suspend a spring from some very secure position, then attach a mass to its free end, the weight of the mass will stretch the spring. If we slowly lower the mass until the spring just holds it still, we have the mass in what we call its *equilibrium* position. If we then pull the mass downward a short distance and release it, it will rise to about an equal distance above its equilibrium position and move downward again to where it was released, keeping up this oscillatory motion, but not moving quite so far each time. Finally, it moves a very small distance up and down about the equilibrium position, eventually coming to rest. If we are going to use amplitude as a variable, we must decide how it is to be measured. As we know, the amplitude of vibration of a spring starts at some initial value Y_0, but then slowly dies out, the mass coming to rest at the equilibrium position after a while. Deciding how to measure the amplitude must be one of our problems in the experimental design.

PROBLEMS

1. A large metal cylinder is supported at its center by a rod of small diameter. The other end of the rod is fixed firmly in a point of support. When the cylinder is twisted from its equilibrium position and then released, it starts to twirl first in one direction; then stopping, it twists back through its equilibrium position until it momentarily stops at some angle of twist in the other direction. It keeps this up until it finally comes to rest again. If the supporting rod stiffness is measured by what is called the *torsion constant*, and the measure of inertia for this system is the *rotational inertia* of the cylinder, what measures of this system are analogous to those of the mass on the spring?

2. List as many vibrating systems as you can, specifying the measure of stiffness, the measure of inertia, and what kind of amplitude the system has.

10-2 *Design of the Experiment*

We have now identified the variables which influence the period of vibration, T, of a specific system—a mass on a spring. They are the mass M, the spring stiffness K, and the amplitude of vibration Y. These variables are ones that our qualitative perception tells us must in some way be responsible for the period T. Stated mathematically, we would write

$$T = F(M,K,Y)$$

This functional equation is a short way of saying "the period of vibration of a mass on a spring depends upon the mass M, the spring constant K, and the amplitude Y".

In designing an experiment to investigate the influence of each of these variables, M, K, and Y, on the period T, we must consider ways of holding all but one independent variable constant, and, while changing that one, observing what happens to the dependent variable T. However, it is not quite this simple. The spring stiffness constant K has been described, but we have not said precisely how it can be measured. Also, the amplitude starts at some initial value and gets smaller, finally becoming zero after some time. What amplitude would we use?

As part of our experimental design, we must decide on a method of measuring each variable. First let us consider the mass M to be suspended from the spring. We shall measure it with a balance, but we want always to have a much larger mass suspended than the mass of the spring. We therefore measure the mass of the spring, and always use a suspended mass which is at least ten times larger.

To measure the stiffness constant K of the spring, we make a series of measurements of displacements. From these measurements, we can determine the weight needed to produce a unit displacement. Let us call this quantity the *spring constant*. Stated mathematically

$$K = \Delta W / \Delta Y$$

the slope of the graph of W on the vertical axis and Y on the horizontal axis. We hope that the graph is a straight line; otherwise, the spring would not be uniformly stiff. We would select springs for which this condition is met. Of course, we are interested in the spring constant only over the distance through which the mass would be vibrating.

The period of vibration T is defined as the length of time it takes the mass to move from its lowest point, past its equilibrium position to its highest point, and back through its equilibrium position to its lowest point again. If you tried to measure the period of a spring with a stop watch, the results would not be too good. The motion happens too fast. We need a way of improving our measurement of T. You probably recognize immediately that we need to let the mass vibrate a large number of times, measuring the total time required and dividing by the number of vibrations to get the period T. Thus, for our measurements of the period, we must count vibrations n and measure a total time t. The period is then a derived measurement given by $T = t/n$. Usually n is taken at about fifty vibrations. To improve our value of T, we could take sets of measurements of 50 vibrations and perform a statistical analysis of these data.

When we measure the period T, the mass is allowed to vibrate 50 times. During this observation the amplitude changes, becoming less and less, but it does not become zero. The mass is still vibrating when we quit timing it. As a measure of amplitude, then, we could use the average displacement of the mass from its equilibrium position. Stated as a formula, this average amplitude, $\overline{Y}$, would be written

$$\overline{Y} = (Y_i + Y_f)/2$$

where Y_i is the initial amplitude and Y_f is the final amplitude.

Now that we have identified each variable and specified how it is to be measured, we can describe an experimental design. We want to investigate the relationship between the period T and the spring constant K, the mass M, and the average amplitude $\overline{Y}$ for a mass suspended by a spring and set into vibration.

The first requirement for this experiment is to find springs. We have selected five springs, each with a different stiffness, that is, a different spring constant K. We have not measured K, but in stretching the springs we notice that they have a different stiffness. Next we select masses to be used. Since the mass used would depend upon how much it would stretch a given spring, we shall start our experiment with a large supply of masses varying in size from 10 g to 500 g. We also need a tally counter and a stop watch to measure the time t for n vibrations, so that we can compute the period. Finally, we need an instrument to measure the initial and final amplitudes of vibration.

To determine the spring constant K for each spring, we must measure weights W and corresponding displacements from equilibrium Y, finding K as the slope of the graph of W on the vertical axis and Y on the horizontal axis. Each spring will have its own graph, with a different slope in each case. These data will supply values for the five constants K_1, K_2, K_3, K_4, and K_5.

We are investigating T as a function of the variables K, M, and $\overline{Y}$. Because K, M, and $\overline{Y}$ are independent variables, we must design our experiment so that two of these three are held constant while the third is varied. We then look for the effect on the period T. The first part of the experiment would involve holding M and $\overline{Y}$ constant, changing K. This means we would use different springs, but always use the same mass and also the same initial displacement from equilibrium.

The second part of the experiment would involve holding K and $\overline{Y}$ constant while we change M. In this case, we would keep K constant by always using the same spring. The average amplitude $\overline{Y}$ would be kept constant by always starting with the same initial displacement. While we change the mass M, we would observe the period T.

In the last part of the experiment, the mass M and the spring constant K would be held constant while the average amplitude $\overline{Y}$ was changed. By using a single spring and a particular mass, both K and

M would be held constant. To change the average amplitude $\overline{Y}$, we could begin vibrations with different values for the initial displacement from equilibrium.

By carrying out the experimental design discussed above, we can secure a set of data which can be analyzed to determine relationships among these variables. The next step is to collect that data.

PROBLEMS

1. To determine the spring constant K for a spring, we find that when 5 lb is added to the weight stretching the spring, the spring is stretched an additional 2 in. What is the value of K?

2. A set of measurements is made to determine the spring constant K of a spring. The total weight W and the corresponding total displacement Y of the spring from equilibrium are shown in the following table. Construct a graph of W on the vertical axis and Y on the horizontal axis, finding K as the slope of this graph.

W (newtons)	Y (centimeters)
100	3.7
200	7.4
300	11.1
400	14.8

3. A mass on a spring is observed to vibrate through one complete cycle 50 times during 80 sec. What is the period T for this system?

4. A mass suspended on a spring has its equilibrium position at the 50 cm mark of a meter stick. When it is displaced to the 60 cm mark and released, the mass vibrates up and down between the 40 and 60 cm marks initially, but after 50 complete cycles it is moving between the 48 and 52 cm marks. What was the initial amplitude? What was the average amplitude?

10-3 *Collection of Data*

In the last section we determined an experimental design. We shall now collect the data according to that design. First, data needed for spring constants will be collected. Next the mass and average amplitude will be held constant while data are collected on the period and spring constant. Then the spring constant and the average amplitude will be held constant while the mass and period are varied. Finally, the mass and spring constant will be held constant while the average amplitude and period are made variables.

To determine the spring constant K, we need measurements of weights needed to extend the spring different distances. For masses, we have convenient sizes of 100, 200, 300, 400 and 500 g. From our previous study of weight and mass, we know these quantities are related by the equation $W = Mg$, where g is the acceleration of gravity. At the location of this experiment, g has a value of 9.80m/sec^2. Expressing M in kilograms and g in these units, our masses have weights of 0.98, 1.96, 2.94, 3.92, and 4.90 newtons respectively. Thus, for example, a 200 g mass has a weight of 1.96 newtons.

Table 10-1 shows data collected for our five springs. Notice that for each spring we have used the same weights, but they produced different extensions for different springs. This difference reflects the fact that the spring constants are different.

TABLE 10-1

	Spring				
	1	2	3	4	5
W (newtons)	Y (m)	Y (m)	Y (m)	Y (m)	Y (m)
0.98	0.024	0.051	0.100	0.031	0.020
1.96	0.051	0.100	0.202	0.068	0.041
2.94	0.074	0.151	0.301	0.100	0.060
3.92	0.101	0.200	0.400	0.134	0.083
4.90	0.123	0.252	0.501	0.169	0.102

For each W measurement, $\delta W = 0.01$ newton.
For each Y measurement, $\delta Y = 0.003$ m.

For the first part of the experiment, the mass M and the average amplitude of vibration $\overline{Y}$ are held constant, while the period of vibration T is measured for different springs, that is, for different spring constants K. To get a more precise value of T, 50 vibrations are counted while a stop watch measures t. The process is repeated for ten trials. Table 10-2 shows the five sets of data, one set for each of the five springs. The mass has been held constant by using 300 g on each spring. The average amplitude has been held constant by always starting the oscillation by displacing the mass from equilibrium a distance of 5.0 cm.

TABLE 10-2

Part I T observed for different values of K

Spring Constant K

Trial	K_1 t (sec)	K_2 t (sec)	K_3 t (sec)	K_4 t (sec)	K_5 t (sec)
1	27.5	39.0	55.0	31.6	24.2
2	27.3	38.6	55.2	32.1	24.6
3	27.6	38.9	54.8	31.8	24.4
4	27.4	38.7	55.0	31.9	24.6
5	27.5	39.1	54.8	32.0	24.4
6	27.7	38.9	55.0	31.7	24.8
7	27.5	38.8	55.2	31.8	24.6
8	27.6	38.8	55.0	32.2	25.0
9	27.4	39.2	55.3	31.5	24.8
10	27.5	39.0	54.7	31.4	24.6

M = 300 g, and constant
$\overline{Y}$ = constant, with Y_i = 5 cm
n = 50 cycles

TABLE 10-3

Part II T observed for different values of M

Mass M

Trial	0.100 kg t (sec)	0.200 kg t (sec)	0.300 kg t (sec)	0.400 kg t (sec)	0.500 kg t (sec)
1	22.4	31.8	39.0	44.8	49.9
2	22.3	31.9	38.6	44.6	50.2
3	22.4	31.8	38.9	45.0	50.3
4	22.5	32.2	38.7	45.2	50.5
5	22.6	31.8	39.1	44.9	50.1
6	22.3	31.9	38.9	44.8	50.2
7	22.4	31.8	38.8	44.9	50.1
8	22.5	31.7	38.8	45.1	50.6
9	22.2	31.7	39.2	44.7	50.3
10	22.4	31.4	39.0	45.0	49.8

$K = K_2$ = constant
$\overline{Y}$ = constant, with Y_i = 5 cm
n = 50 cycles

For the second part of the experiment, the spring constant K and the average amplitude of vibration $\overline{Y}$ are held constant, while the mass

[§ 10-3] Collection of Data 199

M and the period T are changing. To hold K constant, we select spring number two. Again, $\overline{Y}$ is held constant by always starting the oscillation by an equal displacement from equilibrium. Masses from 100 g to 500 g are used, periods of time t for 50 vibrations being observed. Because of variations in these times, ten trials are observed for each of the five masses. Table 10-3 shows these data.

Finally, we must examine the dependence of the period T on the average amplitude $\overline{Y}$. To do this, the mass will be displaced by some initial amplitude; then, when the amplitude decreases to 1 cm, the clock will be stopped and the number of cycles counted. With the initial and final amplitudes known, we can compute the average amplitude from the defining formula. As shown in Table 10-4, the spring constant K and the mass M are held constant. The final amplitude Y_f is also held constant, making the average amplitude depend upon the initial amplitude Y_i, which is changed.

TABLE 10-4

Part III T observed for different values of $\overline{Y}$

Initial amplitude Y_i

Trial	2 cm $n = 34$ t (sec)	5 cm $n = 50$ t (sec)	8 cm $n = 62$ t (sec)	11 cm $n = 75$ t (sec)	14 cm $n = 89$ t (sec)
1	26.4	39.0	48.3	58.3	69.5
2	26.5	38.6	48.2	58.4	69.1
3	26.6	38.9	48.3	58.3	69.0
4	26.5	38.7	48.4	58.2	69.3
5	26.7	39.1	48.2	58.3	69.2
6	26.5	38.9	48.3	58.1	69.4
7	26.3	38.8	48.4	58.2	69.3
8	26.4	38.8	48.3	58.5	68.9
9	26.5	39.2	48.2	58.3	69.1
10	26.6	39.0	48.4	58.4	69.2

$K = K_2 =$ constant
$M = 0.300$ kg $=$ constant
$Y_f = 1.0$ cm $=$ constant

With the data of Table 10-4, the collection process is complete. An experimental design was made, and data collected according to that design. The problem now is to analyze these data, looking for relationships among the variables. This process will be accomplished in the following sections.

PROBLEMS

1. A pendulum swinging back and forth is an oscillatory system. Design an experiment and collect data for the period of a pendulum (one swing over and back) in terms of the length, the mass of the bob, and the amplitude.

2. A wire stretched between two points is an oscillatory system. Stringed musical instruments depend upon this fact. The tension F in a string can be controlled by using weights as shown in the figure below. The amplitude can be measured by observing the deflection during vibration. Design an experiment and prepare tables for collecting data to study the relationship between the period T, the tension F, the length L, and the amplitude of vibration $\overline{Y}$. The period is given by $1/f$, where f is the frequency of vibration. The frequency can be measured by comparing the tone produced by the vibrating wire with one of a set of tuning forks, using your ear as a frequency comparator.

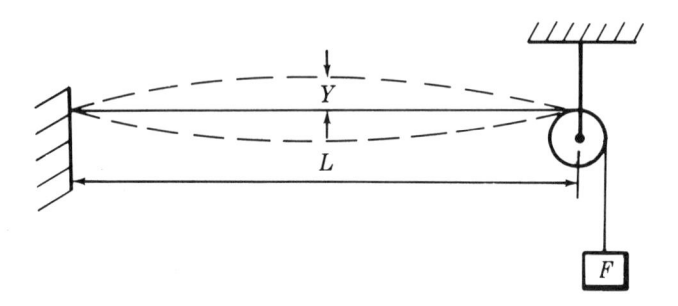

10-4 Data Analysis

An experiment has been designed, data have been collected, and they are ready for analysis. We cannot find a relationship among the variables T, M, K, and $\overline{Y}$ until we perform some sort of data analysis with our measurements in Tables 10-1, 10-2, 10-3, and 10-4. These analyses must involve processes of computation, and, when repetitions were made, statistical analyses must be applied.

The data in Table 10-1 consist of measurements to determine the constants K_1, K_2, K_3, K_4, K_5 for the springs. We know that the spring constant is given by the slope of the graph of the weight W on the vertical axis and the extension Y on the horizontal axis. Figure 10-2 shows the graph for spring number three. Notice that even though each point has a width of 0.006 m, corresponding with 2 δY, and a thickness of 0.02 newtons, corresponding with 2 δW, the points are very small. The relative uncertainties are quite small for the data of this graph. As shown in the figure, the slope is 9.80 newtons/m. This result means that if the spring could be stretched 1 m, 9.80 newtons would be needed to do so.

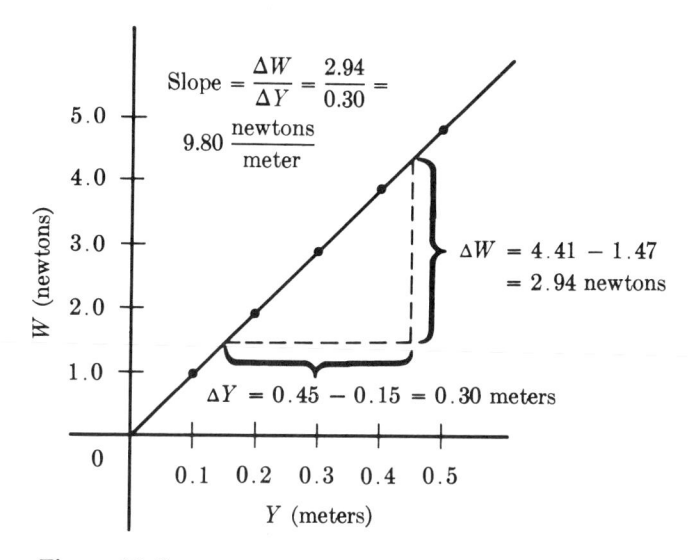

Figure 10-2. *Determining a spring constant from slope.*

When graphs similar to Figure 10-2 are made for the rest of the data in Table 10-1, we get values for the spring constants as shown in Table 10-5. These spring constants will be used as variables when we analyze later experimental data.

TABLE 10-5

Spring Number	Spring constant K (N/m)
1	39.5 ± 0.3
2	19.5 ± 0.1
3	9.80 ± 0.01
4	29.6 ± 0.3
5	48.2 ± 0.2

The uncertainty shown for each spring constant has been determined by substituting values of Y and W from Table 10-1 into the equation $W = KY$, doing a statistical analysis on the five K values. The uncertainty is the computed probable error of the mean in each entry of

Table 10-5. As an example, we found $K = 9.80\ N/m$ from the slope of the graph of Figure 10-2. Because our graph passes through the origin, the equation of it must be $W = KY$. Substituting values of Y and W from Table 10-1 for that spring, using the slope measurement as the mean, we can compute the standard deviation, the standard deviation of the mean, then the probable error of the mean. That result is, as shown in Table 10-5,

$$PE_{\overline{K}} = \delta K = 0.01 N/m$$

Now that the spring constants have been determined, the experimental data can be examined. In each of Parts I, II, and III, the data involved repetitions. We must therefore do a statistical analysis on each column in Tables 10-2, 10-3, and 10-4. These analyses should provide probable errors of the means as absolute uncertainties in the periods. In each table we have 10 trials, therefore $N = 10$. We must compute the mean in each case first; then we must find deviations from the mean, square these, and find sums of squared deviations. With these quantities, we can find the standard deviation, the standard deviation of the mean, and then the probable error of the mean. By assuming that the mean is our best estimate of the period, and the probable error of the mean is our best estimate of the absolute uncertainty in the mean, we can prepare simple tables for the three parts of our experiment. Let us examine only one set of data statistically, giving the results for the rest of the data, but leaving out the computational details.

TABLE 10-6

Trial	t (sec)	$|\overline{t} - t|$	$|\overline{t} - t|^2$
1	31.8	0.0	0.00
2	31.9	0.1	0.01
3	31.8	0.0	0.00
4	32.2	0.4	0.16
5	31.8	0.0	0.00
6	31.9	0.1	0.01
7	31.8	0.0	0.00
8	31.7	0.1	0.01
9	31.7	0.1	0.01
10	31.4	0.4	0.16

$\Sigma t = 318.0 \qquad \Sigma |\overline{t} - t|^2 = 0.36$

$\overline{t} \quad = \Sigma t/N = 318.0/10 = 31.80$

$S \quad = \sqrt{\Sigma |\overline{t} - t|^2/(N-1)} = \sqrt{0.36/9} = \sqrt{0.04} = 0.20$

$S_{\overline{t}} \quad = S/\sqrt{N} = 0.20/\sqrt{10} = 0.20/3.16 = 0.062$

$PE_{\overline{t}} = 0.67\ S_{\overline{t}} = (0.67)(.062) = 0.042$

[§ 10-4] Data Analysis 203

For our example, let us use the data for the 0.200 kg mass in Table 10-3. These data have been reproduced in Table 10-6, where the statistics have also been computed. For these data, we find a mean of $\bar{t} = 31.80$, a standard deviation of $S = 0.20$, a standard deviation of the mean of $S_{\bar{t}} = 0.062$, and a probable error of the mean of $PE_{\bar{t}} = 0.042$. We therefore state the period $T = t/n$ as

$$T = (\bar{t} \pm \delta\bar{t})/n = (31.80 \pm 0.042)/50$$

which is $T = (0.636 \pm 0.001)$ sec.

When these kinds of computations are made for all the data, we can summarize the experimental data in Tables 10-7, 10-8, and 10-9. When these data are analyzed, we can find the relationships among the experimental variables.

TABLE 10-7

T (seconds)	K (newtons/meter)
0.550 ± 0.001	39.5 ± 0.3
0.778 ± 0.001	19.5 ± 0.1
1.100 ± 0.001	9.80 ± 0.01
0.636 ± 0.001	29.6 ± 0.3
0.492 ± 0.001	48.2 ± 0.2

$\bar{Y}$ = constant, M = constant

TABLE 10-8

T (seconds)	M (kilograms)
0.448 ± 0.001	0.100 ± 0.001
0.636 ± 0.001	0.200 ± 0.001
0.778 ± 0.001	0.300 ± 0.001
0.899 ± 0.001	0.400 ± 0.001
1.003 ± 0.001	0.500 ± 0.001

$\bar{Y}$ = constant, K = constant

TABLE 10-9

T (seconds)	$\bar{Y}$ (meters)
0.779 ± 0.001	0.015 ± 0.001
0.778 ± 0.001	0.030 ± 0.001
0.779 ± 0.001	0.045 ± 0.001
0.778 ± 0.001	0.060 ± 0.001
0.778 ± 0.001	0.075 ± 0.001

M = constant, K = constant

PROBLEMS

1. By graphing W on the vertical axis and Y on the horizontal axis using data from Table 10-1, determine from the slope the values of K for springs one, two, four, and five. Compare your results with those listed in Table 10-5.

2. Using data for the 0.400 kg mass in Table 10-3, perform an analysis similar to that done in Table 10-6, finding the mean, standard deviation, standard deviation of the mean, and probable error of the mean. Then using the formula $T = (\bar{t} \pm \overline{\delta t})/n$, find the best estimate of the period T and its absolute uncertainty.

10-5 Experimental Results

The three tables at the end of Section 10-4 summarize the results of this experiment, but tabular entries can be put into more meaningful form as graphs and equations. The techniques learned in Chapter 9 can be applied here to determine the relationship of the period to the other variables.

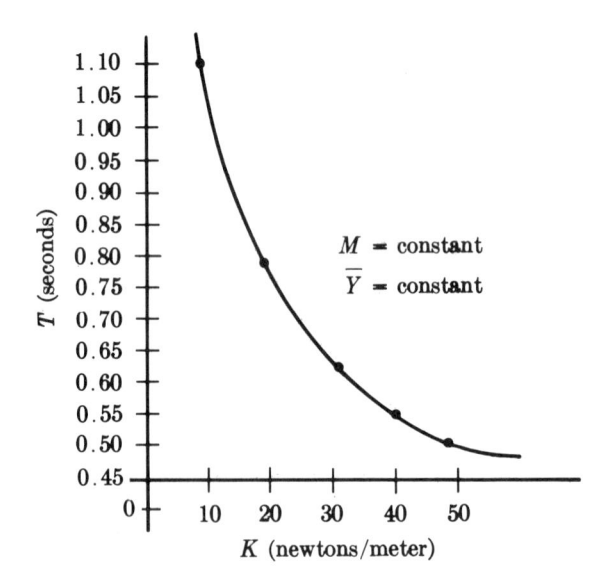

Figure 10-3. *The period graphs non-linearly with K.*

When the pairs of numbers in Table 10-7 are graphed, we are presenting the relationship of the period to the spring constant, other variables

[§ 10-5] Experimental Results 205

being held constant. Figure 10-3 shows this plot on ordinary graph paper. Because we do not get a straight line, the measurements are then graphed on log-log paper in Figure 10-4. The slope is $-1/2$ and the vertical intercept is 3.44. From these results, we can present the relationship by the equation

$$T = (3.44)K^{-1/2}$$

But $K^{-1/2}$ is the exponential form of $1/\sqrt{K}$; therefore, our equation becomes

$$T = 3.44/\sqrt{K}$$

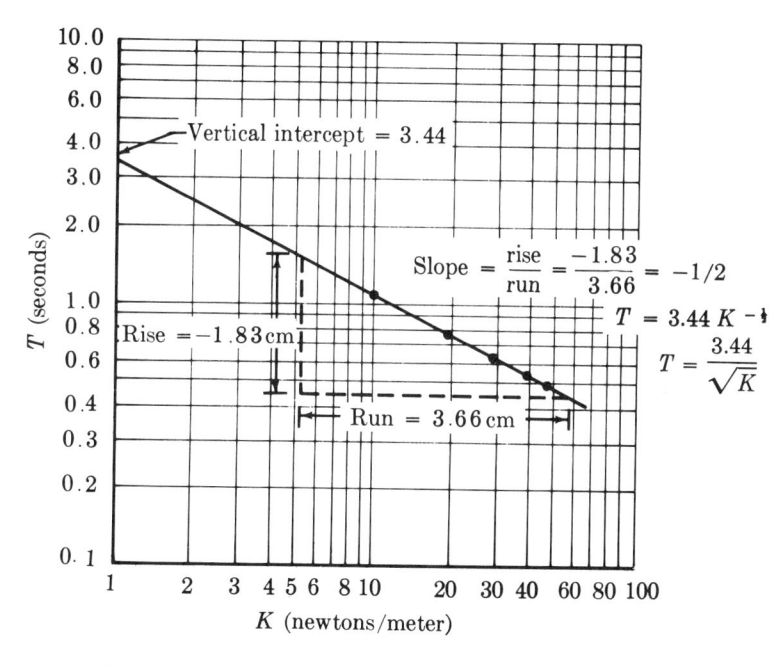

Figure 10-4. *A straight line results on log-log paper.*

When the pairs of measurements in Table 10-8 are plotted on ordinary graph paper, a non-linear curve again appears. When the plot is made on log-log graph paper in Figure 10-6, the slope is $1/2$ and the vertical intercept is 1.425. These results indicate that the correct equation relating these variables is

$$T = (1.425)M^{1/2}$$

Because $M^{1/2}$ is the square root of M, the equation can be written

$$T = (1.425)\sqrt{M}$$

Notice that the vertical intercept is taken where the line crosses the vertical axis at the horizontal coordinate whose logarithm is zero. Because $\log_{10} 1 = 0$, this must be at the number one. Depending upon how the scales are selected, this vertical axis could be at the right side, left side, or even elsewhere on the log-log graph paper.

Finally, we plot a graph of the results in Table 10-9, where M is constant, K is constant, and we are finding the effect of amplitude of vibration on the period. As shown in Figure 10-7, the plot is a straight line on ordinary graph paper, with a slope of zero. From this result, we know that the period of oscillation is independent of the amplitude, at least over the range in which this experiment was conducted.

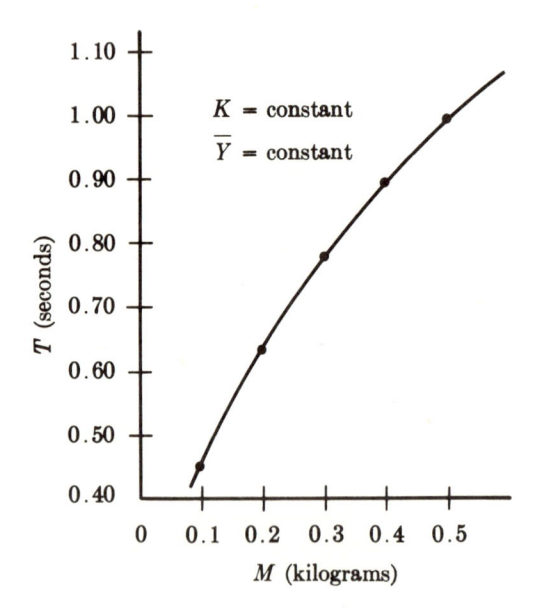

Figure 10-5.   *The period graphs non-linearly with M.*

We have used log-log graph paper, where equations of the form $V = V_I H^S$ graph as straight lines. V is the vertical variable, V_I is the vertical intercept, H is the horizontal variable, and S is the slope. We were thus able to determine that the period of vibration of a mass on a spring followed the relationships $T \sim 1/\sqrt{K}$ with M constant, and $T \sim \sqrt{M}$ with K constant. We found that T was independent of the average amplitude of vibration $\overline{Y}$.

[§ 10-5] *Experimental Results* 207

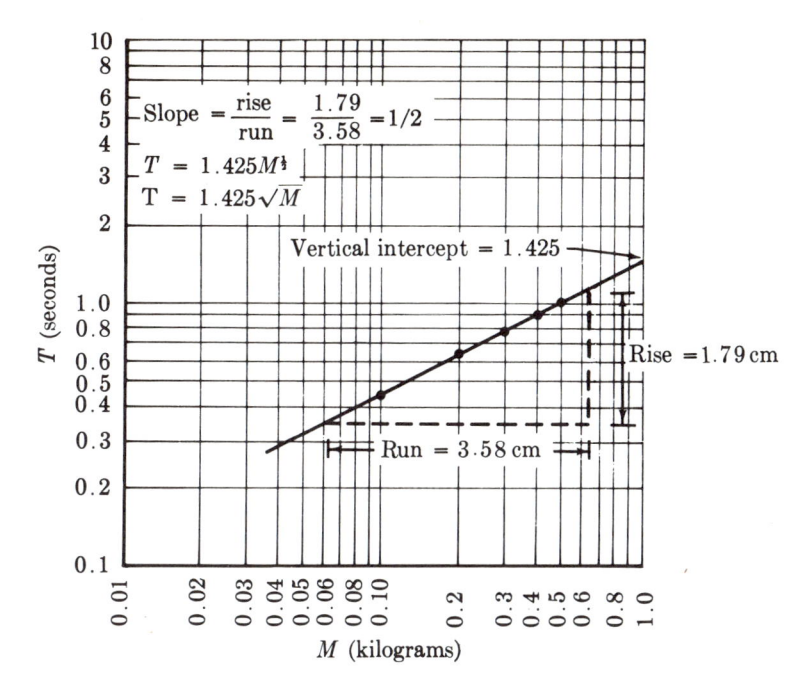

Figure 10-6. *A straight line results on log-log paper.*

Because T is directly proportional to the square root of M and inversely proportional to the square root of K, we can write these two relationships together by the statement $T \sim \sqrt{M/K}$. Using a proportionality constant, C, this proportion becomes the equation

$$T = C\sqrt{M/K}$$

To evaluate the constant C, let us examine the two separate equations and what quantities were held constant.

To get the equation

$$T = 3.44/\sqrt{K}$$

M was held constant at $M = 0.300$ kg. From the equation

$$T = C\sqrt{M/K}$$

the number 3.44 must be the same as

$$C\sqrt{M} = C\sqrt{0.300} = (0.547)C$$

Thus, we get

$$(0.547)C = 3.44 \quad \text{or} \quad C = 6.28$$

To check this result, let us perform a similar examination of the equation

$$T = 1.425\sqrt{M}$$

Again, 1.425 must be the same as $C/\sqrt{K}$. But for that graph data,

$$K = K_2 = 19.5\ N/m$$

We then have

$$1.425 = C/\sqrt{19.6} \quad \text{or} \quad 1.425 = C/4.42$$

and we have $C = 6.28$.

We have just found that the constant in our equation relating T, M, and K has the value $C = 6.28$. We can now write the equation as

$$T = 6.28\sqrt{M/K}$$

This equation summarizes the results of the experiment. It can be used to predict the period of vibration of a mass on a spring. It seems quite remarkable that an experiment as apparently involved as this one should have as a complete result such a simple statement as this equation:

$$T = 6.28\sqrt{M/K}$$

but this is it.

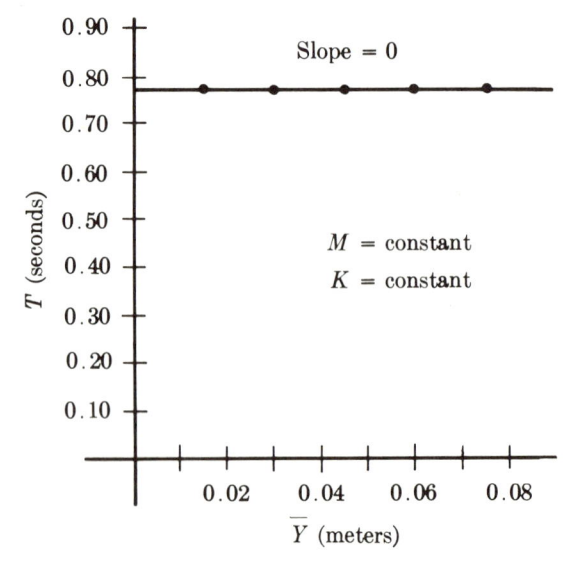

Figure 10-7.   *The period is independent of Y.*

PROBLEMS

1. When an experiment is performed to investigate the relationship of the period T of a pendulum to the length L of the pendulum, the result is $T \sim \sqrt{L}$. A doughnut-shaped satellite in orbit about the earth could be rotated at various rates to produce a force to give objects weight. We could walk around inside such a rotating object, *downward* being directed radially outward. If we change the rate of rotation of the "doughnut", we change the value of g in this system. Thus we could perform experiments to determine the relationship of T to g for a pendulum. When this experiment is performed, the following results are observed.

T (seconds)	g (m/sec/sec)
3.46 ± 0.02	3.30 ± 0.03
2.79 ± 0.02	5.44 ± 0.03
2.13 ± 0.02	8.72 ± 0.03
1.99 ± 0.02	9.94 ± 0.03

   Use log-log paper to determine the formula relationship between T and g. Then combine the above relationship to length with this one for g. How does this result compare with that for a mass on a spring?
   $L = 1.00$ m.

2. When the period of a pendulum is measured with a fixed length $L = 1.00$ m, a fixed value $g = 9.80$ m/sec^2, and the average amplitude as the distance from the vertical axis through the point of support horizontally to the bob, the following data shows the period T as a function of the amplitude $\overline{Y}$. Does the period of a pendulum depend upon the amplitude? If so, is the relationship simple? How does this result compare with that for a mass on a spring?

T (seconds)	$\overline{Y}$ (meters)
2.01 ± 0.01	0.020 ± 0.01
2.01 ± 0.01	0.040 ± 0.01
2.01 ± 0.01	0.080 ± 0.01
2.01 ± 0.01	0.100 ± 0.01
2.01 ± 0.01	0.200 ± 0.01
2.02 ± 0.01	0.300 ± 0.01
2.04 ± 0.01	0.500 ± 0.01
2.13 ± 0.01	0.867 ± 0.01

10-6 Conclusions

This experiment began as an investigation of simple harmonic motion—the vibration of systems. Variables were identified as relating to the stiffness of a vibrating object, the time required for the vibration to complete one cycle, some measure of inertia of the system, and the amount of vibration, called amplitude. By specifying a particular system, a mass on a spring, we were able to determine the relationship $T = 6.28\sqrt{M/K}$, where T is the period, M is the mass suspended, and K is the spring constant.

What has been accomplished in this experiment may not necessarily be applied to other vibrating systems. It has been an *empirical* investigation. We found a relationship for a particular system, but we have not considered at all a theory to explain what we have observed.

The theory of simple harmonic motion uses certain basic laws of physics to analyze oscillatory systems. When these laws are applied to a mass suspended by a spring, the theoretical prediction is that the period should follow the formula

$$T = 2\pi\sqrt{M/K}$$

Remembering that our constant was 6.28, which is approximately equal to 2π, we did get a result quite in agreement with theory.

It is interesting to consider theoretical formulas for the period of oscillation of other systems. For a cylinder supported by a thin rod, the formula is

$$T = 2\pi\sqrt{I/C}$$

where I is the rotational inertia and C is the torsion constant of the rod. For a simple pendulum,

$$T = 2\pi\sqrt{L/g}$$

where L is the length and g is the acceleration of gravity. In a certain tuned electrical circuit,

$$T = 2\pi\sqrt{L/(1/C)}$$

where L is the inductance of the circuit and $1/C$ is the reciprocal of the capacitance of the circuit. It is interesting that the inductance L offers opposition to changes in electrical current; it has a sort of inertia. Reciprocal capacitance is very much like a stiffness constant. This electrical circuit is an electrical analog of the mechanical system. Even though the units of L and $1/C$ are quite different from M and K, the units of $L/(1/C)$ are identical with those for M/K.

This experiment has led to the discovery of an empirical law. We also know that theory predicts the same law. This has been but one kind of experimental investigation. Sometimes a theory will make a prediction about some system that no one has ever looked for before.

Because the theory makes a prediction, we often investigate these predictions to test our theory. If the prediction is right, we feel better about the theory, but it does not prove the theory correct. If the prediction is wrong, however, and the experiment has been performed carefully by a competent person, the theory must be rejected or modified. The scientist's ultimate appeal is to nature herself.

PROBLEMS

1. Electrical inductance L is measured in units called *henrys*, capacitance in units called *farads*. We want to design an electrical circuit which has the same period of oscillation as a 0.400 kg mass on a spring of $K = 43$ N/m. If we have available only an 800×10^{-6} f capacitor, what size must we use for an inductance?

2. The formula for the period of vibration of a weight on a spring in the English system is given by

$$T = 2\pi\sqrt{(W/g)/K}$$

where W is the weight in pounds, K is the spring constant in lb/ft, and g is the acceleration of gravity, 32 ft/sec/sec. A 4000 lb automobile having no shock absorbers is observed to hit a bump and oscillate up and down on its springs at a rate of 5 cycles in 4.0 sec. Assuming that 1/4th the weight of the car is on each spring, what is the spring constant?

11 Reading Technical Materials

Scientific and technical reading materials are written quantitatively. They use words having specific meanings. Symbols are often used to make principles and concepts very concise. Graphs, figures, and diagrams are placed in the text to replace "a thousand words". Reading such material must be done differently from reading other non-technical books.

When you are reading a novel, skipping a few pages or a paragraph here or there does not alter the overall impression conveyed. Quickly skimming through a novel at 1500 words per minute permits you to follow the story being told. However, scientific and technical books must be read slowly in some parts, faster in others, but always with care that full meaning is taken from the printed page. Just "saying" the words, ignoring figures, diagrams, and graphs, is not reading! You must refer to illustrations as they occur. You must follow derivations carefully, understanding details of steps taken, and filling in those left out by the author. Only if you gain full meaning from what is said, done, and shown, do you "read" these kinds of materials.

11-1 Reading Definitions

In studying a foreign language, you would not think it unusual to memorize definitions of several thousand words. In a science or engineering course, where only a few definitions are needed, people often do not even try to memorize definitions. Yet without a complete understanding of fundamental definitions, it is impossible to comprehend the principles and concepts of any technical subject. Definitions form the foundation upon which the entire structure is built.

Why do people apparently ignore definitions? It is partly because they do not recognize definitions in their reading. We do not like to stop reading long enough to memorize definitions as they occur. Such reluctance to take time to memorize definitions in the course of reading is false economy. It would not be long before ignorance of the definition would prevent one from understanding the text.

One of the major reasons people fail to memorize definitions is their inability to distinguish a definition from a law or principle. Most technical definitions are best made using mathematical equations. Laws and principles are also stated in this form. Unless the author has made it quite clear when he is stating a definition, the reader is likely to confuse it with a law. It is not uncommon for someone to want to know how to "prove" a definition. A definition is made in order to describe some property of physical quantities. Definitions are arbitrary. Once stated, with all terms clearly defined, we agree to use the definition with that particular meaning. No one ever needs to prove a definition, nor could he do so!

How then do we recognize a definition when it is made? Authors are seldom consistent in the way definitions are expressed, but there are certain words that indicate a definition is being made. For example, look at the following definitions.

"*Nonconcurrent forces* are forces whose lines of action do not intersect at a common point."

"A single force which will just cancel the effects of several forces is called the *equilibrant*."

"The *angle of repose* is defined as the steepest slope angle for which an object will remain at rest on a given inclined surface."

"The ratio of useful work or work output to the work input is the *efficiency*."

"The process of finding the resultant is described as the *composition of vectors*."

"Whenever the resultant force system acting upon an object is zero, the object is in *equilibrium*."

There is one characteristic almost all word definitions have in common. The quantity, process, condition, or concept being defined is stated in italics. This procedure is followed by most authors. It is something to look for in reading. All italicized words are not definitions, but almost all word definitions are italicized.

The next property of definitions to look for is a set of key words. Of course, the words "defined as", or any expression using "defined", or "definition" is a clear indication of a definition. Less obvious words involve some statement followed by "is called the", "is described as", or even, "is", and then the italicized word which is being defined. Sometimes the italicized word is preceded by a set of conditions, as in the last example above. Being aware of these indicators will help you recognize definitions as they occur in your reading.

Because formula definitions are less common to the average person, they seem to cause the greatest difficulty. The indicators just discussed are still helpful in recognizing such definitions. However, instead of merely defining the quantity in words, the formula definition describes the quantity in terms of certain mathematical operations to be performed on other previously-defined quantities. Look at the following examples.

"The *work* W done on an object by a constant force F, where the object is moved a distance D in the direction of the force, is given by the equation $W = FD$."

"The equation $E_k = (1/2)MV^2$, where M is the mass, and V is the speed, defines *kinetic energy*."

"The mass per unit of volume, $\rho = M/V$, is called *density*."

"The *amplification factor* α is the slope of the graph of collector current I_c as a function of emitter current I_e, that is $\alpha = \Delta I_c/\Delta I_e$."

These are definitions. The words, "is", "is called", "is defined by" are indicators of that fact. Again, the italics designate the quantity being defined. However, the definitions are in terms of mathematical operations like multiplication, division, squaring, and so forth. To understand one of these definitions, it should be used in a numerical example.

We have examined various kinds of definitions and the ways that you can recognize a definition. Now, what do you do when a definition occurs in your reading? First, stop and examine it carefully. Be certain that you understand all words used in the definition. If you do not, go back to where the word was first used, and learn its meaning. If it is a formula definition, you must understand what each quantity in the definition represents. Sometimes this understanding depends upon knowing how each quantity was defined earlier in your text.

In a definition, a mathematical operation could be used that might be unfamiliar to you. If this occurs, you must find out what the operation means. For example, in the definition of the amplification factor of a transistor, the delta symbols may not be understood; or, a more complicated and general definition of *work*, $W = \int \vec{F} \cdot d\vec{r}$, would be very difficult for one to understand without a knowledge of calculus. To understand a formula definition, use numbers to work out a specific example. Usually, examples are given in texts for important definitions, but you should make your own when they are not present.

When you understand what a definition means, then memorize it. You cannot prove a definition. You cannot derive it. A definition may not make any apparent sense in other respects. It is introduced for convenience. The discussion which will follow it depends upon its being understood and remembered. If you take notes as you read, write the definition in the notebook, expanding any words unfamiliar to you, or defining any quantities not already familiar. Then stop reading long enough to memorize the definition. This task can usually be accomplished in only a few minutes by looking away and trying to recall the definition, referring to your notes only when you cannot recall some part. When the definition can be recalled without difficulty, reading may be continued. The delay will seem troublesome, but the gains in understanding, and the net result in learning, will make your effort worthwhile.

PROBLEMS

1. A mass of 4 kg is moving at a speed of 15 m/sec. What is its *kinetic energy*?

2. The work output of a machine is 50 ft-lb. If the input work is 85 ft-lb, what is the *efficiency* of the machine?

3. The kinetic energy density is defined by the kinetic energy per unit of volume. Make a formula definition for kinetic energy density in terms of mass M, speed V, and volume. Notice that we cannot use V for volume, since we have already used V for speed. Use some other symbol for the volume.

4. By constructing a table of numbers and graphing them, make an example for computing the amplification factor of a transistor. Then memorize the definition, timing with a watch how long it takes to memorize it.

11-2 Graphs, Figures and Diagrams

Most technical and scientific books are filled with illustrations. The illustration may be a simplified representation of some arrangement of apparatus being discussed. It may be a graph of the relationship of two variables. It may be just a photograph. Whatever is being depicted, or whatever the method used, an illustration is there for a purpose. It should be referred to and examined as directed in the textual material. Illustrations are not put into books just to fill up space so that you do not have to read as many words. They are there to help you understand what is being discussed.

We have already discussed in detail how to construct graphs of various kinds from tables of numbers. When you know how it is made, reading a graph or interpreting it in terms of the slope or intercept is simple. The graph is used in reading materials to illustrate a relationship. Some of these relationships could be shown with an equation, but a graph presents a more meaningful expression, enabling you to "see" how one variable changes with changes in another. Consider the following example of the use of a graph in a physics textbook.

"Real gases, unlike ideal gases, have intermolecular forces. At pressures and temperatures where the molecules of the gas are brought close to each other, there will be a departure from Boyle's Law. ...Figure... shows a plot of pressure versus volume for different temperatures of a real gas, ... Curves A and B, ... are hyperbolas conforming to Boyle's Law. In curve C we see a departure from Boyle's Law."*

The graph is referred to in several different ways. Unless you stop reading after each referral to the graph and examine that part of it, you cannot understand the discussion. Sometimes references are made to points on graphs. Others may be to the shape of the curve, or the slope at some point. But whatever is said, you must examine the illustration and note the characteristic being discussed. It is part of reading quantitative materials.

Figures illustrating arrangements of apparatus and the physical quantities involved are often combined. Such figures simplify the apparatus parts to idealized objects without supporting structure. In

---

*Alexander Joseph, *et al.*, *Physics for Engineering Technology* (New York: John Wiley & Sons, Inc., 1966), p. 422.

[§ 11-2] *Graphs, Figures and Diagrams* 217

Chapter 10 of this book, Figure 10-1 showed a mass suspended on a spring. The figure illustrated a mass and a spring, but it did not look like a real spring or a particular mass. The discussion referred to the arrangement of mass and spring.

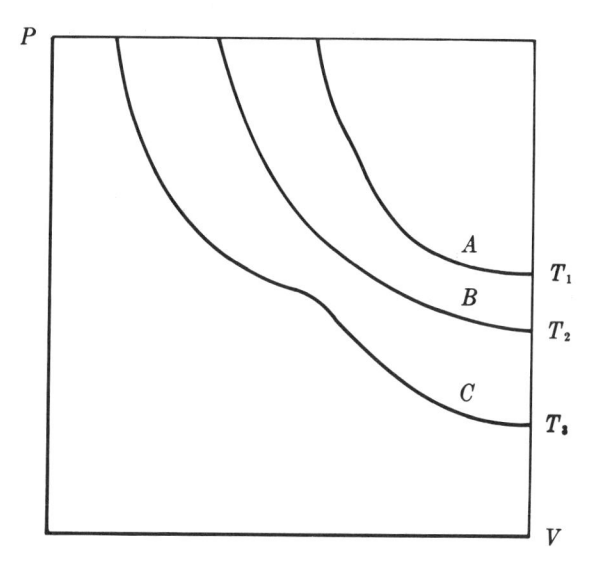

Figure 11-1. *A graph.* *

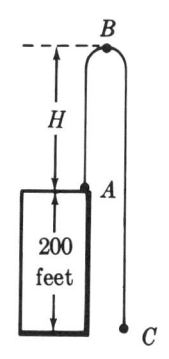

Figure 11-2. *The path is better shown with a figure.*

Some descriptions of situations are so complicated that a figure is essential. For example, try to read the following passage without a

---

Ibid., p. 423.

figure. An object is thrown vertically upward with some initial speed, V_0. It travels a distance H, stops, reverses its direction, and comes back down the distance H plus and additional 200 ft.

Now read the same passage with references to Figure 11-2. An object is thrown vertically upward from point A with an initial speed of V_0. It travels a distance H to point B, where it stops, reverses direction, and comes back down, passing its position at A and continuing an additional 200 ft to point C.

Some illustrations are complicated by including arrows which represent directed physical quantities in the same figure with objects and the arrangement of apparatus. These figures permit discussing the spatial arrangement of the objects and the directions of the physical quantities more easily. Figure 11-3 shows a point mass moving at a speed V in a circle of radius R. Using this illustration, we could discuss an important derivation in mechanics, the formula for the centripetal acceleration.

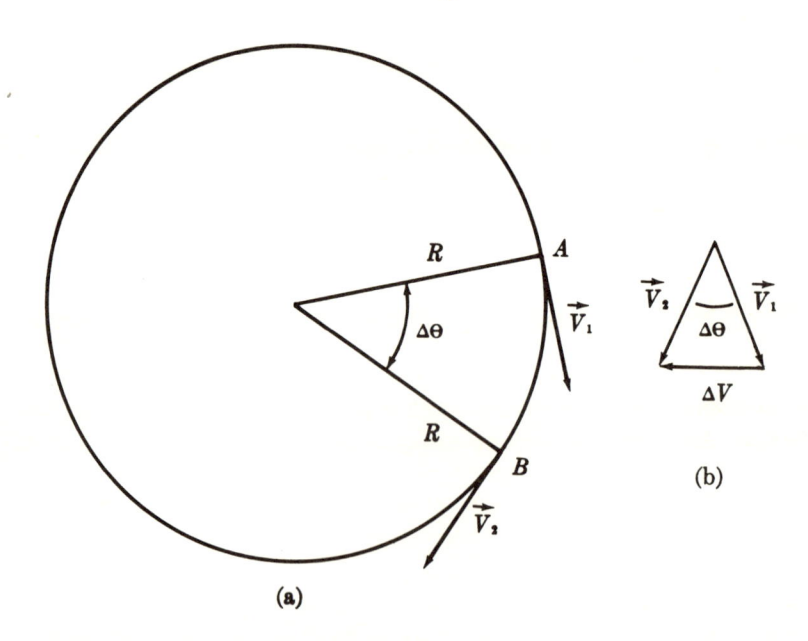

Figure 11-3. *A figure used in a derivation of centripetal acceleration.*

We have examined a few ways in which figures, graphs and diagrams are used in scientific and technical materials. These examples indicate the importance of referring to illustrations in technical reading. It should be expected to slow you down, but remember that comprehension is your primary concern. Most technical illustrations are an essential part of the reading that accompanies them.

PROBLEMS

1. Convert the following statement into a diagram. An arrangement of pulleys supports a mass M. A rope passes under the first pulley which is fastened to a table top. The rope then goes to the ceiling, passing over a pulley fixed to the ceiling, and back down to the mass.

2. Read the following passage. A point mass moves in a circle of Radius R at a constant speed V. Its velocity at point A is shown as $\vec{V}_1$, and at point B by $\vec{V}_2$. Between vectors $\vec{V}_1$ and $\vec{V}_2$ we note that the angle $\Delta\theta$ is the same as made by the radii to points A and B since the vector velocities each make $90°$ angles with the radii. Now read the passage again, this time referring to Figure 11-3 (a) and (b). Why are the angles equal as described?

11-3 Deductions and Derivations

When you have learned to stop reading to memorize definitions as they occur, and when you have developed the ability to examine illustrations carefully as technical materials are read, your comprehension will increase substantially. However, there is another reading skill you must develop: the ability to read *derivations* and *deductions*.

What is a derivation or deduction? It is a logical process, usually in a mathematical form, leading from a set of hypotheses to some conclusion. As a high school geometry student, you performed a large number of derivations. In geometry, they are called *proofs*. In algebra, a derivation might be a process of performing operations to both sides of an equation to change the form of the equation. It might be a process of starting with two different equations involving the same variables, but ending with an equation that tells you more than the separate equations apparently did. In technical subjects, a derivation usually starts with separate laws, principles, and definitions. The use of mathematical operations leads to some result not apparent in the separate equations.

Reading derivations requires that you follow each step carefully. An author seldom includes every possible step in a given derivation. He usually includes a minimum number according to the level of his readers. For example, in an advanced physics textbook the reader might be expected to fill in a page or so of handwritten steps to get from one equation to the next in some derivation. It is expected in these cases that the reader has at some time performed these steps in a similar mathematical problem. Unfortunately, this is not always true, and a graduate student might spend several hours "reading" a three line derivation on a page of a book.

For most elementary science or engineering derivations, usually only a few steps are left out. Even then, the reason for going from one step to another is seldom given. You should be certain that you understand why one part of a derivation follows from another. You should keep notes as you read, carrying out such derivations in their entirety, filling in whatever steps are necessary to *your* understanding of the derivation. It is not so important that you remember every step of all derivations studied as it is that you remember how they are made.

Let us consider as an example a derivation involving certain kinematics formulas. These formulas describe the motion of objects. As symbols, we use V_f for the final speed, V_0 for initial speed, a for acceleration, X for distance traveled, and T for time required. By using fundamental definitions of kinematics, the following formulas can be deduced:

$$V_f = V_0 + aT \qquad (11\text{-}1)$$

$$X = V_0 T + (1/2)aT^2 \qquad (11\text{-}2)$$

A textbook might use these two equations as hypotheses to deduce another equation. Depending upon the level of the book, steps in that derivation might be omitted. A text could state, "from Equations (11-1) and (11-2), we see that $V_f^2 = V_0^2 + 2aX$". How you would go from those two equations to this one may not be at all clear to you.

In a less difficult passage, the text would read, "By solving Equation (11-1) for T, and substituting the result into Equation (11-2), we have $V_f^2 = V_0^2 + 2aX$." Even though this passage tells you a little more about how to go from the two given equations to the result, the details of the process are not entirely clear. You would be expected to figure this out for yourself, filling in the missing steps. Certainly you should not just accept the result, without showing that it does follow from the two original equations.

Let us look now at the way this derivation would probably be stated in an elementary technical book. It would begin, "solving Equation (11-1) for T, we have

$$T = (V_f - V_0)/a \qquad (11\text{-}3)$$

It would then say, "substituting this value of T into Equation (11-2), we have

$$X = V_0[(V_f - V_0)/a] + (1/2)a\,[(V_f - V_0)/a]^2 \qquad (11\text{-}4)$$

Finally, "squaring the quantity in the second term of Equation (11-4), adding the two terms, and solving the result for V_f^2, we have

$$V_f^2 = V_0^2 + 2aX \qquad (11\text{-}5)$$

It is doubtful that a text would give more steps than have been included here, yet this still leaves much to be done. Figure 11-4 shows the kinds of notes which should be taken while "reading" this derivation. Notice how all steps have been filled in completely. Later, you can refer to these notes when reviewing the material. You understand how Equation (11-5) follows from Equations (11-1) and (11-2) because you made the deduction yourself. The text has just guided your thinking. It told you how the derivation could be made, but you made the derivation.

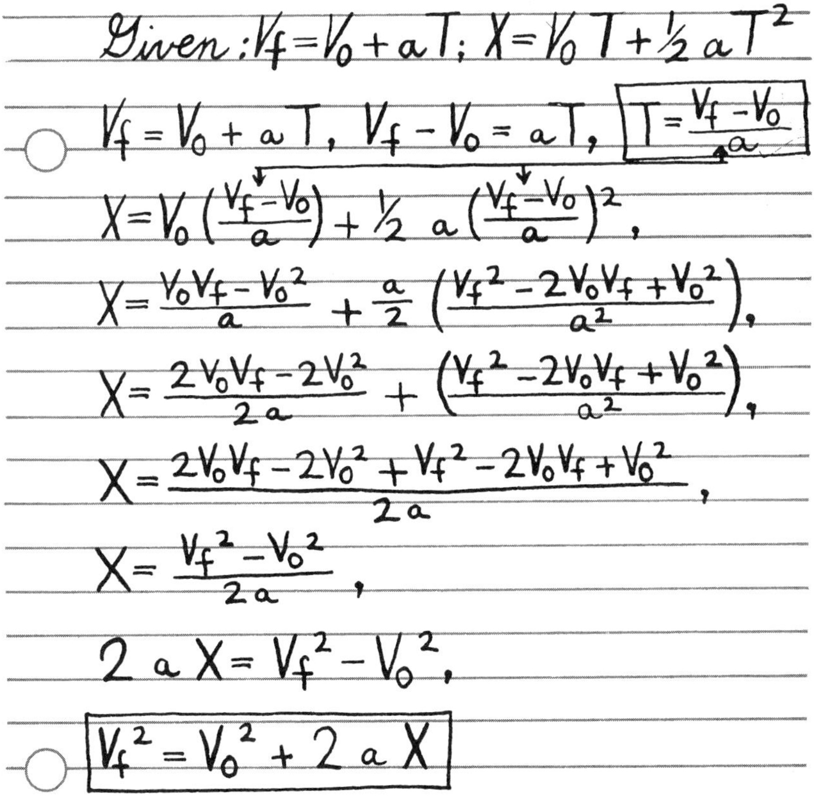

$$\text{Given}: V_f = V_0 + aT; \quad X = V_0 T + \tfrac{1}{2} a T^2$$

$$V_f = V_0 + aT, \quad V_f - V_0 = aT, \quad \boxed{T = \frac{V_f - V_0}{a}}$$

$$X = V_0 \left(\frac{V_f - V_0}{a}\right) + \tfrac{1}{2} a \left(\frac{V_f - V_0}{a}\right)^2,$$

$$X = \frac{V_0 V_f - V_0^2}{a} + \frac{a}{2}\left(\frac{V_f^2 - 2V_0 V_f + V_0^2}{a^2}\right),$$

$$X = \frac{2V_0 V_f - 2V_0^2}{2a} + \left(\frac{V_f^2 - 2V_0 V_f + V_0^2}{a^2}\right),$$

$$X = \frac{2V_0 V_f - 2V_0^2 + V_f^2 - 2V_0 V_f + V_0^2}{2a},$$

$$X = \frac{V_f^2 - V_0^2}{2a},$$

$$2aX = V_f^2 - V_0^2,$$

$$\boxed{V_f^2 = V_0^2 + 2aX}$$

Figure 11-4. *A notebook page showing a derivation.*

How do you recognize derivations or deductions when they are presented in a text? They usually involve words like, "substituting for ... we get ...", or, "from Equation ... we have ...", or, "Equation ... then becomes ...", etc. A statement is usually made about some physical system. Then certain fundamental laws are presented which apply to that system. Finally, it is stated that some other equation or result follows from what has been said. Of course it does not follow unless you have taken the time to perform the mathematical steps in your notes to get the stated result. Only then can you understand how the result follows, and what, if any, assumptions were made in getting the result.

The importance of assumptions can be illustrated by considering another example from elementary kinematics. Imagine that you accept the formula $D = 16T^2$ for how far, D, an object falls in some time T. If you just memorize the formula, without having derived it yourself, you will be unaware of the assumptions used in its derivation. You may therefore use it incorrectly. For example, it is valid only when the initial speed is zero. The coefficient, 16, is only approximately correct as one half the acceleration of gravity in the English system. This quantity varies slightly over the surface of the earth. On the moon, this expression would be completely incorrect. The equation ignores air resistance; therefore, it can apply at best to very dense masses which do not fall very far, since the equation is approximately valid only near the surface of the earth.

Derivations and deductions in technical books do not include all steps needed to make the processes understood. If you are to understand these important reading materials, you must keep notes as you read, filling in missing steps, and not reading further until you fully understand a given derivation. When you have reached advanced courses and the author states that a result follows from some given equations, if you had at some time in your past experience performed identical operations to get results like these, you can just recall the method used and not actually carry out every detail. You learn to skip steps which are simple for your level of mathematical development. But never skip a step you do not understand.

PROBLEMS

1. For the following derivation, fill in the missing steps. The general gas law states that $PV = nRT$, where P is the pressure, V is the volume, n is the number of moles of the gas, T is the absolute temperature, and R is a constant. We know also that $m = nM$ where m is the total mass of the gas and M is its molecular mass. From these two equations and the definition of density, $m/V = \rho$, we can find an expression for the density of a gas in terms of its pressure, temperature, and molecular

mass. Solving $m = nM$ for n and substituting the result into $PV = nRT$, we have $PV/RT = m/M$. Then, multiplying both sides of this equation by M and dividing both sides by V, we have

$$PM/RT = m/V \quad \text{or} \quad \rho = PM/RT$$

2. Perform the missing steps in the following derivations. Ohm's Law states that $V = IR$, and the power formula gives $P = IV$; it then follows that $P = I^2R$. We can also show that $P = V^2/R$.

3. Carry out the missing steps.
   a. For a pendulum of small amplitude,

   $$T = 2\pi\sqrt{L/g}$$

   therefore, solving for g, we have

   $$g = 4\pi^2 L/T^2$$

   b. Three capacitors are connected in parallel across a battery of voltage V. This means that V is placed across each capacitor and the charge on each capacitor is given by $Q = CV$. Because the total charge must be

   $$Q_T = Q_1 + Q_2 + Q_3$$

   we must have

   $$Q_T = C_1 V + C_2 V + C_3 V$$

   But since $Q_T = C_T V$, we get

   $$C_T V = C_1 V + C_2 V + C_3 V \quad \text{or} \quad C_T = C_1 + C_2 + C_3$$

   as a formula for the total capacitance of three parallel capacitors.

11-4 How to Study an Assignment

We have discussed some of the techniques of reading quantitative materials. Definitions must be understood and memorized as they occur. Figures, diagrams, and illustrations form an essential part of a technical book; they must be referred to and examined as part of the reading process. Derivations and deductions must have missing steps filled in by the reader in the notes he would take to accompany his reading.

The reading skills discussed here are quite different from those needed in reading a novel, or in studying non-quantitative textbook materials. Such reading is necessarily variable in speed, some parts requiring much time and others, primarily verbal passages, taking less. A person must develop an ability to be flexible in his reading rate.

He must aim for complete comprehension, and he must evaluate it continuously.

When you have learned how to read scientific and technical materials, there are still certain study techniques that can help you become an efficient reader. When you are given a reading assignment, the previous assignment should be reviewed before starting the new one. Then the present assignment should be scanned quickly. Finally, the assignment should be read, where "read" means to take full meaning from the printed page. If problems are part of the assignment, you may want to look at them to see what you must be able to do with what you are reading, but you should not attempt to work such problems until you have read the assignment.

Topics in science or technology are almost always built one upon another in a logical fashion. When you are given a specific assignment, say to read three sections of your text and work problems at the ends of these sections, that assignment is not independent of everything else you have studied and what you will be studying. The topics covered undoubtedly depend upon what you have previously studied. You should therefore review quickly the previous assignment, summarizing that material and perhaps listing important equations, laws or formulas from it.

By scanning the assigned reading, you can gain some idea of where you are going, of what will be covered. This process gives you a sense of direction. It will indicate the extent of the assignment and of the kinds of coverage, quantitative, descriptive, verbal, or mathematical. Scanning the assignment prepares you for the intensive study which should follow. Looking over the problems gives you direction concerning what kinds of comprehension is expected. It points out what you should be able to do with the assigned concepts, laws, and principles.

The next step in doing an assignment involves reading and taking notes. Do not even start a reading assignment until you are prepared to take notes as you read. Each time a definition appears, whether it be verbal or a formula definition, write it down, understand it, and memorize it before continuing. When the text refers to a figure, diagram, or illustration, look at it, examining each part of the figure as directed in the reading. Make your own diagram or figure in your notebook, adding to it if necessary. Be certain to follow derivations and deductions carefully. Steps omitted in the text should be filled in completely in your notes. When you cannot understand how the author got from one step to the next, make a note of this fact. This question will be welcomed by your instructor, since it shows that you made a genuine effort to understand the derivation before asking for an explanation.

When the assignment has been read carefully, and notes taken, they should be summarized. The important definitions, laws, principles, and concepts should be stated briefly at the end of the detailed notes for this assignment. This summary can then be used as a review when beginning the next assignment.

Only after the reading assignment has been studied by the methods detailed above should the assigned problems be worked. You will find that it will be much easier to tackle the problems with this preparation.

Almost everyone has difficulty with problems in science and technical textbooks. It is usually because of so-called "word problems". These kinds of problems require that the reader structure them—set them up for solution. He has to know how to analyze a problem, select relationships needed, make derivations or deductions, construct diagrams, and make computations. There are specific techniques which can be used to work such problems. Those techniques will be discussed in the following chapter.

Let us summarize the steps you would take in studying a reading and problem assignment.

1. Review the previous assignment.
2. Scan the present reading assignment, examining figures, diagrams, illustrations, and assigned problems.
3. Read the assigned topics, taking careful notes on derivations and definitions, referring as directed to figures, graphs, and illustrations.
4. Summarize the assignment in your notebook, listing important laws, principles, definitions, and results of derivations.
5. Work assigned problems, using the techniques to be developed in the next chapter.

PROBLEMS

1. Let us say that you have been given an assignment to read Section 8-5 of this book. Follow the steps suggested at the end of this section, doing step number one, reviewing the previous assignment, by reading Section 8-4.

2. Read each of the following textbook sections, taking notes and following the suggested study procedures outlined in this chapter.
   a. Section 2-1   b. Section 5-4   c. Section 6-4

12 Solving Technical Problems

If there is one difficulty encountered by most people in science, engineering, or mathematics courses, it is working what are called "word problems". These problems approach real situations by including verbal descriptions with references to numerical values of quantities. Word problems involve mathematical operations, but they require considerably more than subcerebral mathematical manipulations.

It is unfortunate that in recent years mathematics courses provide little experience with word problems. Mathematics instruction, although aiming sincerely at more profound understanding, has tended to omit applications in word problems. This omission is partly due to the difficulty of working such problems, but it is also partly due to a feeling among mathematicians that such material tends to contaminate the "purity" of mathematics. In spite of efforts to the contrary, students are gaining from their mathematics courses abilities to manipulate symbols, but few analytical skills.

[§ 12-1] *Reading a Problem* **227**

There are certain techniques which can assist you in attacking word problems. These techniques include a careful reading and dissection of the problem to find out what is *given* and what is *wanted*. When you have this ability, you must then draw on your experience with the subject to relate this problem to that material. Then you can solve the problem.

12-1 Reading a Problem

To understand how to work a word problem, you must understand what a word problem is. A technical book is attempting to teach you about real situations in science or technology. The fundamental laws and principles of science are applied in the factories and laboratories. To gain some insight into how these laws can be applied, or to enhance your understanding of some principle, a nearly real situation is presented to you in a text; then some problem is established for you to solve.

Basically, a word problem has two parts. It describes a physical system, or establishes some situation; then it asks you to find out something about this system or situation. These two parts of the word problem could be symbolized by the words, *given* and *find*. *Given* would include certain measurements. It would describe the physical system; it would establish the situation. *Find* would be the part of the problem asking you to do something.

Reading a word problem is not different from reading text materials in science or technology. Usually, however, the reader must provide his own figure or diagram. He must re-structure the problem until he completely understands what is being said. Only then should he examine the problem to find out what is wanted as a solution.

Let us examine a word problem of the simplest form. "A room is 10 ft wide and 12 ft long. What is the area of the room?" To read this problem, we must find out what is given and what we are to find. We could draw a diagram, but the problem is really too simple for that. We are given $W = 10$ ft and $L = 12$ ft. We are asked to find the area, $A = ?$. These things seem obvious, but the process of putting the problem into quantitative form is important in getting it ready for solution. The reading of this problem would be summarized on your paper by the following:

Given	Find	Variable definition
$L = 12$ ft	$A = ?$	L: length of room
$W = 10$ ft		W: width of room
		A: area of room

We shall now consider a somewhat more difficult problem. "A 1000 kg car is moving along a level roadway at a speed of 25 m/sec when the brakes are applied in such a way as to produce constant deceleration. If the car travels 90 m along the road before coming to rest, what is the force acting on the car while it is being stopped?" What is given in this problem? Reading the first sentence, we should stop each time some quantity is mentioned. The statement "A 1000 kg car" tells us that $M = 1000$ kg. We should write that on our work sheet. Next, the problem says "is moving along a level roadway at a speed of 25 m/sec". That the car is moving on a level roadway tells you it will not be going downhill or uphill, something which might complicate the problem. That it is moving at 25 m/sec can be stated by $V_0 = 25$ m/sec. We use the zero subscript on the V to indicate that this is an initial speed. The statement "to produce a constant deceleration" tells us that this negative acceleration is constant, and we could write a = constant. In the next sentence, "the car travels 90 m along the road" can be summarized by $X = 90$ m. Finally, the words "before coming to rest" tell us that for the final speed $V_f = 0$.

To determine what is wanted, we look for a question. In this problem, it is easy to find: "What is the force acting on the car while it is being stopped?" We can write that question as $F = ?$. On our work sheet, the reading of this problem would appear in the following way.

Given	Find	Variable definition
$M = 1000$ kg	$F = ?$	M: mass of car
$V_0 = 25$ m/sec		V_0: initial speed of car
$X = 90$ m		V_f: final speed of car
$V_f = 0$ m/sec		F: force stopping car
		X: distance car moves while stopping

A problem of this sort can be read a little better if accompanied by a sketch of a graph. What we do is make a visual picture of the motion of the car by plotting a graph of the speed as a function of time. As shown in Figure 12-1, the area under such a curve gives the distance traveled. The slope of this curve is the acceleration of the car. Since the slope is negative, the car is decelerating as described. These kinds of graphs, quickly sketched out, are particularly helpful in solving problems involving motion. But they may also be useful for other technical word problems.

As a final example, let us read the following problem. A mass of 300 g is suspended from a spring. When displaced from its equilibrium

[§ 12-1]   Reading a Problem   229

position and then released, it oscillates at a frequency of 5 cycles/sec. What would be the frequency of oscillation of a 2 kg mass if it were suspended from this same spring and set into harmonic motion?

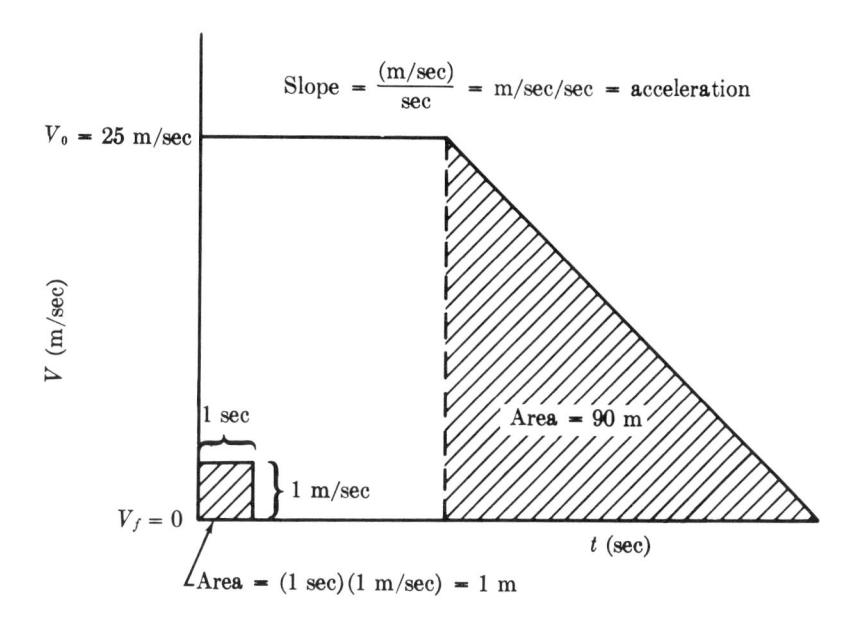

Figure 12-1. *A graph shows the motion of the car.*

Reading the problem through tells us that there are two different circumstances or situations to consider, one case when the 300 g mass is used, the other case when the 2 kg mass is used. Thus we must designate our information accordingly. For given information, we have in Case I, $M_1 = 0.300$ kg, $f_1 = 5$ cycles/sec, and we are using a spring in simple harmonic motion. Notice that we have changed 300 g to 0.300 kg. This procedure should always be followed in a problem, since formulas are in terms of fundamental physical quantities. In the MKS system, those quantities are meters, kilograms, and seconds. The fact that we have a spring suggests an additional piece of given information, namely, there is some spring constant K_1.

In Case II, the given information becomes $M_2 = 2.0$ kg, and we are using the same spring, that is, $K_2 = K_1$. The only quantities which change are the mass and the frequency. It is of course made clear that we are trying to find the frequency in Case II. The reading of this problem would be noted as shown on page 230.

230   Quantitative Aspects of Science and Technology   [§ 12-1]

Case	Given	Find	Variable definition
I	$M_1 = 0.3$ kg $K_1 =$ constant $f_1 = 5$ cycles/sec		M: mass on a spring K: spring constant f: frequency of oscillation of mass on spring
II	$M_2 = 2.0$ kg $K_2 = K_1 =$ constant	$f_2 = ?$	

These examples illustrate the basic technique of reading a word problem. The method is similar to reading other parts of a quantitative textbook. Specific steps are as follows.

1. Scan the problem quickly. This means to read the words of the problem, getting a general idea of what you will be doing to work it.
2. Carefully analyze the words making up the problem, searching for given information. Such information should be written on a work sheet, indicating each bit of information with a clearly defined symbol. Whenever possible, a figure, diagram, or graph should be constructed showing how the various bits of given information are related to each other.
3. Read the problem to determine what is wanted. This task usually involves looking for a question being asked or something to be done by the reader. You must clearly define the quantity to be found, marking it on your work sheet.

When these three steps have been accomplished, you are ready to begin working the problem. Usually, this initial process is the most difficult part of working word problems. It is the shock of seeing a long passage of facts, measurements, and descriptions that causes the most difficulty. So much at one time appears confusing. Only when the problem is dissected, as we have shown here, does it become sufficiently less imposing that a solution can quickly follow. In the following sections, we shall see how problems of various kinds can be solved.

PROBLEMS

1. Read the following problems carefully, identifying given information. Use equations as in examples of this section to indicate such given information. Define each symbol carefully. When a diagram, figure, or graph could be helpful in presenting given information, construct it.
    a. A wheel is 50 in. in diameter. A steel ring, with a diameter 0.05 in. undersize at 20° C is to be shrunk on. To what temperature must it be heated to make the diameter 0.05 in. oversize?

b. The induction coil used to produce the spark in an automobile spark plug consists of a primary coil, and a secondary coil of many more turns of wire, both wound around a straight iron core. The primary current is broken every time a spark is desired. If the primary winding has 300 turns and normally has flowing in it a current of 5 amp, which falls to zero in 0.003 sec when the circuit is broken, what must be the mutual inductance if an average of 18,000 v is to be induced in the secondary?

2. In each of the following problems, read the problem, showing the given information, an appropriate figure, diagram, or graph to support it, and what is to be found. Define each variable used, and follow the note format used in this section, specifically, Given, Find, and Variable definition.
  a. A car weighing 32 lb rolls from rest at the top of a hill 64 ft high. The slope down which the car moves is inclined at an angle of 30° with the horizontal. What is the initial potential energy of the car?
  b. An 8 kg mass is connected to a 4 kg mass with a string that passes over a pulley of mass 6 kg, radius of gyration 9 cm, and radius 13 cm. Assuming no friction and no slipping of the string, find the acceleration of each mass and the tension in each part of the string.

12-2 Selecting Relationships for Problem Solution

The initial task of reading a problem to determine what information is given and what is to be found leaves one with specific quantities, some known, others not. These quantities must in some way be related to each other, or else the problem could not be solved. The next step in solving a word problem involves finding the appropriate relationship.

The dissection and analysis of a problem leads to given quantities and a quantity which is to be found. If we are to find the unknown quantity, we must find a relationship that contains only the quantities we know and the one we don't. How do we find that relationship? We search our mind to see if we can recall a formula, law, definition, or principle that fits this requirement. If we can't, then we look back over our reading of the assignment preceding the problem to find the relationship. The problems are usually used to exemplify some of the formulas developed in that preceding section. Sometimes a problem will include a law developed earlier, but if required, it would probably be so fundamental that you would recognize the need for it immediately.

In the first example discussed in Section 12-1 we had $L = 12$ ft and $W = 10$ ft for given information, and we were to find the area, A. You can probably recall immediately the appropriate relationship. We need a formula that has in it length L, width W, and area A, with the figure being a rectangle. You remember the formula for the area of a rectangle as $A = LW$. Here we know two of the three variables, and we are trying to find the third. We have found the correct relationship for the problem.

The next example in that section is somewhat more complicated. The reading analysis gave the following:

Given	Find	Variable definition
$M = 1000$ kg	$F = ?$	M: mass of car
$V_0 = 25$ m/sec		V_0: initial speed of car
$X = 90$ m		V_f: final speed of car
$V_f = 0$ m/sec		F: force stopping car
		X: distance car moves while stopping

To find a relationship for this problem, we would have to locate a law, principle, definition, or formula that contained M, V_0, V_f, X, and F, and only these variables. The correct formula in this case is

$$F = M(V_f^2 - V_0^2)/2X \qquad (\textbf{12-1})$$

Notice that the equation contains only the variables mentioned, and it is already solved for the desired unknown.

In the last example of Section 12-1 we had a spring on which we placed two different masses, measuring the frequency of oscillation in the first case. We wanted to know the frequency for the second mass placed in oscillation on the same spring. We must find a relationship which contains frequency f, mass M, and spring constant K, and no other variables. Referring to our reading notes, we would find it:

$$f = (1/2\pi)\sqrt{K/M} \qquad (\textbf{12-2})$$

Even though this equation contains quantities like $1/2$ and π, the only variables are f, K, and M.

These examples should illustrate the task of searching your mind, your notes, or your text for a relationship appropriate to a given problem. The main point to remember is that the relationship must contain as variables only the given quantities and the quantity to be found. It may also contain known constants like π or other physical constants which you may find in tables.

Unfortunately, many relationships needed to solve problems are not so readily available as has been indicated here. The correct formula containing just the given quantities and the unknown quantity may have to be derived. You may have to start with more fundamental laws and definitions and establish this new formula yourself. In the

[§ 12-3]   *Deducing Needed Relationships*   **233**

following section, we shall examine methods of performing such necessary derivations.

PROBLEMS

1. Your notes contain the following relationships involving distance X, time T, force F, Mass M, acceleration A, initial speed V_i and final speed V_f: $F = MA$; $V_f^2 = V_i^2 + 2AX$; $A = (V_f - V_i)/T$; $X = V_iT + (1/2)AT^2$; $V_f = V_i + AT$. In each of the following problems, determine what is given, what is to be found, and select one of these relationships appropriate to the problem.
    a. A force of 30 newtons acts on a mass, producing an acceleration of 5 m/sec/sec. What is the mass?
    b. An object is projected with an initial speed of 14 m/sec. It rises until it stops after 1.5 sec. What was its acceleration?
    c. How far does an object travel in 10 sec when it is accelerating at 8 m/sec/sec and has a beginning speed of 4 m/sec?

2. Your notes contain the following relationships: $V = IR$; $P = I^2R$; $P = IV$; $f = (1/2\pi)\sqrt{1/LC}$; $P = V^2/R$, where V is voltage, I is electric current, R is resistance, f is frequency, L is inductance, C is capacitance, and P is power. For each of the following problems determine what is given, what is wanted, and select the appropriate relationship from this list.
    a. A tuned circuit has an inductance of 4 h and a capacitance of 9×10^{-6} f. What is the frequency of this circuit?
    b. How much power is dissipated in a circuit having a resistance of 12 Ω and a current of 9 amp?
    c. A resistor of 83 Ω dissipates 400 W of power. What voltage is applied across the resistor?

12-3 *Deducing Needed Relationships*

The kinds of problems discussed so far in this chapter are what we call "plug-in type problems". They require a minimum amount of thought, and are therefore considered the lowest level of word problems. Most good word problems must approach more realistically the kinds of situations a technician or scientist would face in practice. Such problems are not so clearly defined as we have indicated here so far.

It is common to have more information given than is actually needed. In being given extraneous measurements, you are facing a problem more like what you have in a real situation, in which you must decide which of many possible measurements to make. You must, therefore, know somehow which variables are important. In a good word problem, you must decide which variables are related and which are merely extraneous. Then you select a relationship which includes those variables.

Plug-in problems have another common feature. They usually require relationships easily found in your notes or in the text. A problem more commonly given requires that you derive the needed relationship from two or more other relationships. There are many formulas possible from the principles of science and technology. All of them cannot be derived in a given textbook. Many of the problems you work therefore are derivations, where you are not expected to use any numbers. The entire problem consists of taking two or more relationships and deducing another relationship. In many problems, where that newly-deduced formula is relatively simple, you are expected to apply it, plugging in quantities for a numerical solution.

As an example, let us go back to Equation (12-1). This equation would most likely be derived from more fundamental equations. The two formulas which lead to this one are

$$F = MA \quad \text{and} \quad V_f^2 = V_0^2 + 2AX$$

A person familiar with mechanics could recall both of these formulas, and he would recognize that when M, V_0, X, and V_f were given and F was to be found, he would need to combine these two equations to get an appropriate formula. Solving the second equation for A, we have

$$A = (V_f^2 - V_0^2)/2X$$

Then substituting this value of A into the equation $F = MA$, we have

$$F = M(V_f^2 - V_0^2)/2X$$

which is Equation (12-1). We have derived an equation to be used in working a problem.

This process of deducing needed relationships is certainly more difficult than solving mere plug-in problems. But it tests your understanding of the concepts and principles better, since it forces you to use those principles in abstract reasoning. You still have a technique for handling such problems. If you are given information for which you have no relationship, use your knowledge of the laws and principles to decide which ones apply. Then select two equations which contain the variables of importance and another variable. One of the equations can then be solved for this unknown quantity, the result being substituted into the second. In this way you will have derived a new formula, as we did to get Equation (12-1).

As a final example, consider the following problem. The voltage across a resistor is 90 v. The current through the 10 Ω resistor flows

[§ 12-3]   *Deducing Needed Relationships*   **235**

for 12 sec. What is the power dissipated? Let us suppose that the only relationships you have available are $V = IR$, $P = IV$, and $Q = CV$, where V is voltage, I is current, R is resistance, P is power, Q is charge, and C is capacitance. The given information in our problems is $V = 90$ v, $R = 10$ Ω, and $T = 12$ sec. We are asked to find the power, P. From your knowledge of electricity, you would know that the time T is not a factor in this problem. Examining the relationships, you see that no one of them will solve the problem. You must derive a formula yourself.

Both R and V are in the formula $V = IR$, and both P and V are in the formula $P = IV$. To get a formula having only P, R, and V, we must solve one of these equations for I, substituting the result in the other. Solving $V = IR$ for I, we get $I = V/R$. Substituting this value of I into $P = IV$, the result is

$$P = (V/R)V = V^2/R$$

We have derived the formula $P = V^2/R$.

We have considered methods of analyzing a word problem to determine given information and what is to be found. With this information, you have learned how to search for appropriate relationships or deduce them as needed. Unless the problem is only a derivation, you still must do a numerical computation. Values of the various known quantities must be placed in the selected relationship, then indicated operations must be performed until you have the value of the unknown. The computational aspect of problem solving will be considered in the next section.

PROBLEMS

1. You are given the following relationships: $X = VT$, $V = (V_f + V_i)/2$ and $V_f = V_i + AT$, where X is distance, V is average speed, T is time, V_f is final speed, V_i is initial speed, and A is acceleration. You need a formula for distance X in terms of initial speed V_i, acceleration A, and time T. Derive that formula by replacing V in the first equation by the right side of the second. Then use the third formula to replace V_f. When you simplify the results of these steps, you will have derived the desired formula.

2. You are given the following relationships: $Q = CV$, and $E = (1/2)Q^2/C$. Read the following problem, indicating given information and what is to be found. Deduce the relationship needed to complete a solution. A capacitor is charged by a voltage of 100 v. If the capacitance is 4×10^{-6} coulombs, how much energy is stored in it?

12-4 Computation in Problem Solving

You have learned to analyze word problems to identify given quantities and what is to be found. You have also studied examples which indicate how to select appropriate relationships for problems. In those cases where no single formula is available, you have seen how derivations produce the needed relationships. The only aspect of problem solving not yet considered is the process of getting a numerical solution, of computation.

Basically, computation is the easiest part of problem solution. For the simplest word problems, you merely plug in the numbers, and "crank out the answers", to use the tritest phrase you will hear in quantitative courses. The first example discussed in Section 12-1 was a problem of this type. You were given $L = 12$ ft and $W = 10$ ft. You knew the formula for area as $A = LW$, thus you plug in the numbers, (12 ft) (10 ft) $= A$, or $A = 120$ ft^2. The numerical part of this computation is trivial.

There are certain characteristics of computation that do deserve consideration here. One of these involves units of physical quantities and how units are treated in computations. The other characteristic concerns uncertainties and how they are propagated through a computation. The purely arithmetic aspect of computation can be performed most of the time with a slide rule, using powers of ten, as developed earlier in this book.

Units of physical quantities should always be included in computation. Indicated mathematical operations should be performed on units, just as they are on the magnitudes of the quantities. If this procedure is followed, the units in your answer should be correct for the quantity being found. Sometimes, however, derived units must be changed to fundamental ones to get a final unit that is recognizable.

The problem of uncertainties is a difficult one. It is hard enough to get a realistic estimate of the uncertainty in a single measurement, much less consider how such errors are propagated through a computation. There are theories of errors based on calculus, but we cannot consider material at that level in this book. There are certain rules, based on theory, that give good estimates of the uncertainty in a computed result. We shall merely state those rules. They are as follows.

Indicated Operation	Rule
Addition or subtraction	Add absolute uncertainties
Multiplication or division	Add relative uncertainties
Powers or roots, $X^{n/m}$	Multiply relative uncertainty by exponent n/m

If these rules are used, a good estimate of the uncertainty of a computed result can be made. If you learn calculus, you can develop more specific formulas for the propagation of uncertainties, but using these rules is better than ignoring the effects of errors. Any error estimate is at best just that, an estimate.

In computation, you will hear people talk about *significant figures*. Simply, significant figures are the numbers of units and subunits *actually counted* in a measurement. Zeros used for place holders at either end of a measurement are not significant figures. This rule indicates that your uncertainty is determined by one half of the last digit indicated. It is analogous to using one half the smallest scale division as the instrument precision. The easiest way to count significant figures is to write the number in power of ten form, counting the number of digits used as a factor times the power of ten. These digits represent the counted numbers of units and subunits, down to where the lack of instrument precision prevented further measurement. Consideration of significant figures gives a rough estimate of the uncertainty in a computed result. You can never have more significant figures in a result involving multiplication or division than you had in the *least* significant factor or divisor.

Let us examine a problem where we carry out the computation, assuming we have determined given information, what is to be found, and the proper relationship to be used. The formula is

$$F = M(V_f^2 - V_0^2)/2Y \qquad (12\text{-}3)$$

We are given that $M = 9.11 \times 10^{-31}$ kg, $V_f = 2.03 \times 10^5$ m/sec, $V_0 = 7.8 \times 10^4$ m/sec, and $Y = 2.143 \times 10^{-3}$ m. This problem is for the force on an electron between two electrodes in a vacuum tube. If no uncertainties are given, you must assume that for each measurement the absolute uncertainty is one-half the last digit. We shall make that assumption in this problem.

Let us make the computation. Substituting in the various quantities, we have

$$F = \frac{(9.11 \times 10^{-31} \text{kg})[(2.03 \times 10^5 \text{ m/sec})^2 - (7.8 \times 10^4 \text{ m/sec})^2]}{(2.000)(2.143 \times 10^{-3} \text{m})}$$

Notice that we have written the pure number 2 in the denominator with as many "significant figures" as any of the factors or divisors. Actually, any such pure number has an infinite number of significant figures.

Let us first carry out the indicated operations on the units. We have

$$F \text{ (units)} = (\text{kg})(\text{m}^2/\text{sec}^2)/\text{m} = (\text{kg})(\text{m})/\text{sec}^2$$

But this is just the unit called a newton in the MKS system. The newton is a unit of force, therefore our formula at least gives correct units in the result. Because we subtracted squared quantities in m²/sec², you might think we should get zero units for the difference. But even though the magnitude of the difference could be zero, the units would remain. In other words, we could have 0 m²/sec² for that difference.

Now we can do the computation. Carrying out the squaring of terms so indicated and performing the subtraction, we have,

$$F = \frac{(9.11 \times 10^{-31})(3.51 \times 10^{10}) \text{ newtons}}{(2.000)(2.143 \times 10^{-3})}$$

Next, grouping powers of ten together, and factors and divisors together,

$$F = \frac{(9.11)(3.51)(10^{-31})(10^{10}) \text{ newtons}}{(2.000)(2.143)(10^{-3})}$$

Finally, doing the multiplication and division, and using rules of exponents on the powers of ten, we get

$$F = 7.46 \times 10^{-18} \text{ newtons}$$

Notice that, in terms of significant figures, 9.11×10^{-31} has three, 3.51×10^{10} has three, and 2.143×10^{-3} has four. Our answer contains only three significant figures because this is the *least* in any factor or divisor.

To find the relative uncertainty in F, we must use the rules previously given for the propagation of uncertainties. Assuming that each absolute uncertainty is one-half the last digit, we get the following absolute and relative uncertainties:

$\delta M = 0.005 \times 10^{-31}$ kg $\delta M/M = 5.5 \times 10^{-4}$

$\delta V_f = 0.005 \times 10^5$ m/sec $\delta V_f/V_f = 2.46 \times 10^{-3}$

$\delta V_0 = 0.05 \times 10^4$ m/sec $\delta V_0/V_0 = 6.31 \times 10^{-3}$

$\delta Y = 0.0005 \times 10^{-3}$ m $\delta Y/Y = 2.33 \times 10^{-4}$

The first indicated operation is squaring the speeds V_f and V_0. We must therefore multiply each of those relative uncertaintities by two. Next we would subtract the squared quantities, so we add absolute uncertainties in the squares of the speeds. Since the relative uncertainty in the squares of these speeds is twice the relative uncertainty in the quantities themselves, the formula for relative uncertainty $\delta V^2/V^2$ tells us to multiply the doubled uncertainty by V^2 to get the absolute uncertainty in V^2. That is, if $\delta V^2/V^2 = R$, then $\delta V^2 = V^2 R$.

Performing these steps on the subtracted squared speeds, we get

$$\delta V_f^2 = 20.0 \times 10^7 \quad \text{and} \quad \delta V_0^2 = 7.68 \times 10^7$$

[§ 12-4]   *Computation in Problem Solving*   **239**

Then adding these absolute uncertainties gives

$$\delta(V_f^2 - V_0^2) = 27.7 \times 10^7$$

for the absolute uncertainty of the difference, and for the relative uncertainty of that difference,

$$\delta(V_f^2 - V_0^2)/(V_f^2 - V_0^2) = 7.9 \times 10^{-3}$$

The rest of our propagation of uncertainties task is somewhat easier, since all quantities are multiplied or divided, and all are to the first power. We need only add the relative uncertainties of the mass M, the squared speed differences, and the distance Y. Adding those relative uncertainties, we get a total of 8.7×10^{-3} for the relative uncertainty in our result. We have, therefore,

$$\delta F/F = 8.7 \times 10^{-3}$$

and an absolute uncertainty of

$$\delta F = (8.7 \times 10^{-3})(7.46 \times 10^{-18}) = 6.5 \times 10^{-20} \text{ newtons}$$

stated differently, we could write

$$F = (7.46 \pm 0.06) \times 10^{-18} \text{ newtons}$$

The problem example just worked illustrates the basic techniques of computation in problem solving. It is painfully clear that the most difficult part of the computation is in the propagation of uncertainties. In most technical courses, you are not expected to perform such analyses on textbook problems. The simple rule of significant figures used appropriately is an adequate consideration of uncertainties. But when you are performing an experiment in the laboratory, where the errors may be very important, the propagation of uncertainties must be carried out carefully.

We had examined methods for analyzing problems to isolate given quantities and what is to be found. You had learned to select or derive appropriate relationships to be used to get numerical answers. In this section you learned to carry out computations, accounting for units and uncertainties in measured quantities.

PROBLEMS

For problems 1 and 2 use $V = (8.0 \pm 0.2)$ v, $I = (3.11 \pm 0.05)$ amp, and $R = (2.6 \pm 0.1)$ Ω.

1.   Using the formula $P = I^2R$, compute P and its absolute uncertainty.

2. Using the formula $P = V^2/R$, compute P and its absolute uncertainty.

3. Using the formula $F = GM_1M_2/R^2$, where $G = 6.67 \times 10^{-11}$ (newton) (m^2)/kg^2, $M_1 = 5.9 \times 10^{24}$ kg, $M_2 = 8.023$ kg, and $R = 6.4 \times 10^6$ m, compute F. Show operations on units, and use significant figures to determine how many digits to retain in your answer.

12-5 Systematic Procedure for Problem Solution

This chapter has been concerned with teaching you how to work scientific and technical word problems. Let us list the steps you would follow in working such problems.
1. Scan the problem quickly to gain some feeling for what it is about.
2. Read the problem carefully, small bits at a time, identifying specifically each given quantity and the quantity to be found.
3. Decide which information given to you is extraneous and not needed. The ability to do this is dependent upon your understanding of the principles, laws or definitions involved. Thus identify the pertinent variables.
4. Search your mind, your notes, or your text to find a relationship which contains only the variables specified in step 3. If you cannot find any such relationship, you must find two or more relationships which do contain these and other quantities. Then deduce the needed relationship yourself.
5. If the problem requires a numerical solution, do this arithmetic, carrying out all operations on the units, and making some estimate of the absolute uncertainty in the computed result. Most often this error estimate can be specified by just considering significant figures.

If the student has followed the suggestions made in Chapter 11 on how to read his text, he will have little difficulty using these techniques in working problem assignments. As a final example, let us work the following problem. A 150 g arrow moving horizontally eastward with a speed of 25 m/sec strikes a 3 kg snark sitting on a tree limb. The arrow becomes embedded in the snark. If the branch on which the snark was sitting is 10 m above the ground, how far eastward will the snark move before "biting the dust"?

Figure 12-2 shows how this problem would be illustrated on a worksheet. The diagram shows the *before* and *after* conditions, listing the given information and the quantity D to be found.

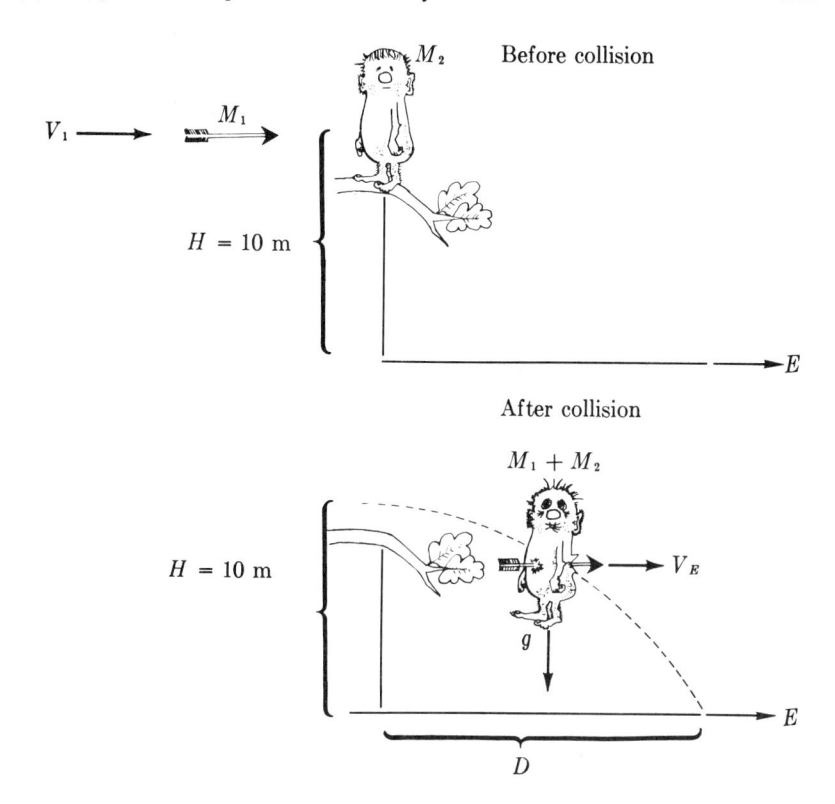

[§ 12-5]   *Systematic Procedure for Problem Solution*   **241**

Before collision

$V_1 \longrightarrow$ $M_1 \longrightarrow$ M_2

$H = 10$ m

$\longrightarrow E$

After collision

$M_1 + M_2$

$H = 10$ m $\longrightarrow V_E$

g

$\longrightarrow E$

D

Given	Find	Variable definition
$V_1 = 25$ m/sec	$D = ?$	V_1: initial speed of arrow
$M_1 = 0.150$ kg		M_1: mass of arrow
$M_2 = 3.0$ kg		M_2: mass of snark
$H = 10.0$ m		H: initial distance of snark above ground
		D: eastward distance snark moves while falling
		V_E: speed of snark in eastward direction

Figure 12-2.   *A work sheet showing given information on a problem.*

A student familiar with laws of physics would recognize that this problem requires the law of conservation of momentum and two kinematics formulas. His relationships would be,

$$M_1 V_1 = (M_1 + M_2) V_E \qquad (12\text{-}4)$$

$$H = (1/2)gT^2 \qquad (12\text{-}5)$$

$$D = V_E T \qquad (12\text{-}6)$$

He must use these formulas to derive a single equation which has in it only the given quantities and the quantity to be found.

Solving Equation (12-4) for V_E, we have

$$V_E = M_1 V_1 / (M_1 + M_2)$$

Substituting this result into Equation (12-6), we get

$$D = T M_1 V_1 / (M_1 + M_2)$$

Finally, solving Equation (12-5) for T gives

$$T = \sqrt{2H/g}$$

and substituting this value of T into our result for D, our derived formula becomes

$$D = (\sqrt{2H/g}) M_1 V_1 / (M_1 + M_2) \qquad (12\text{-}7)$$

To find out how far the snark and arrow move eastward, we must now substitute quantities into this derived equation.

Using the fact that g is a constant equal to 9.8 m/sec², our units for D become

$$D \text{ (units)} = \sqrt{[m/(m/sec^2)]} (kg)(m/sec)/(kg)$$

which simplifies to meters, as expected.

When the arithmetic is done, we get

$$D = (\sqrt{(2)(10)/9.8})(0.150)(25)/3.15$$

which is $D = 1.70$ m. In terms of significant figures, we would probably retain only two digits, giving $D = 1.7$ m. This problem illustrates most of the techniques needed to work problems in science and technology.

We have examined ways to approach problems, and certain techniques for carrying out their solution, but there is no general prescription for handling every situation. Indeed, if there were, you would probably not be interested in science or technology. It is this great variety of phenomena based on a few basic laws, and the challenge of careful and logical thought about them, that make these subjects so interesting.

PROBLEMS

1. Using the five suggested steps for problem solution mentioned at the beginning of this section, and given that $U = wh$ as a relationship where U is potential energy, w is weight, and h is height in feet, work Problem 2a at the end of Section 12-1.

2. In the example problem developed in this section, replace the 150 g mass by a mass of 20 g, the speed of the arrow by 200 m/sec, and the 3 kg mass by 1 kg. Then carry out every step of the solution, including an appropriate diagram, using the example as a guide.

13 Sources of Technical Information

The vast acceleration of science and technology in the world has produced changes to which mankind has had difficulty adjusting. We are so accustomed to living in this complex world that we fail to appreciate the recency of most changes. Yet it is startling to realize that fully 90 per cent of the scientists in the history of the world are alive today. Most of what has happened, especially in technology, has occurred in the past 100 years.

A major factor in our great advances in science and technology is communication among people working in these areas. This communication has been possible because of one thing—printing. The scientific and technical journal has spread knowledge rapidly throughout the world. When reports stood the tests of verification and confirmation from other competent persons, the principle, concept, or law was incorporated into textbooks so that dissemination could be extensive.

13-1 Textbooks

You probably feel that you are all too familiar with textbooks. You do not need to be told about them. But there are some things about textbooks with which you may not be familiar. An understanding of these aspects may give you a greater appreciation for what a textbook is, what its limitations are, what to expect from it, and how to use it.

When the scientific or technical community develops a set of concepts, definitions, principles, or laws in a given area, usually through considerable trial and error effort, and often in a very haphazard way, this body of knowledge is organized carefully, explained, and put into a textbook. In this way, people can learn quickly and efficiently the essence of that particular subject at the time the book was written. Knowledge developed over several hundred years may be expressed here in a beautiful, logical, efficient form.

It is in some ways unfortunate that a textbook is logical, efficient, and beautiful. It tends to make the reader feel that scientists and engineers think in the same logical, efficient and beautiful ways. The reader finds he has so much difficulty assimilating all this knowledge that he thinks there is something wrong with him. The scientific and technical journals are filled with poor procedures, incorrect conclusions, wrong assumptions, dead-end theoretical discussions, etc. The scientist makes many mistakes in his search for understanding. He just does not put them in textbooks, for fear of producing confusion. All of this talk may not make reading a text any easier for you, but it should encourage you to persist. Do not let textbook reading difficulties diminish your determination to understand the subject. The text is the best, most efficient device we have for presenting and explaining science and technology.

Basically, a textbook is written to convey to you an understanding of a certain set of concepts, principles, laws, and definitions. The author begins with these items and attempts to explain them, illustrate them, and give you problems to work to establish your understanding of them.

A textbook represents the state of some science at the time it is written. Except for very fundamental sciences, where major laws have changed very little in the past 50 years, minor revisions are sufficient to keep books on these subjects up-to-date. Often, however, new revisions are made to better explain topics in such books, or to make the examples and applications more modern. For technical books, changes are taking place so fast that completely new textbooks are needed every few years. Certainly major revisions are called for in almost all such books regularly, especially at the advanced level.

13-2 Parts of a Book

It is remarkable that some people know so little about books. Their reluctance to use textbooks properly can be related to their ignorance of the parts of a book. Yet, with knowledge of these parts, you can use books more efficiently and rapidly. Such knowledge enhances your skill in studying.

What are the parts of a book? They are the *title page*, the *preface* or *foreword*, *contents*, *illustrations*, *introduction*, *text*, *glossary*, *appendix*, *bibliography*, and *index*. Any given book will have whatever parts that are actually needed. The book you are reading now has no *introduction*, *glossary*, *bibliography*, or *appendix*.

The *title page* serves to name the work by a title. It may include a subtitle which describes the title better or clarifies it. The author's name and facts about his status appear next. Such information as his position, academic degrees, or titles of other books may appear here. If there is an editor, illustrator, or if someone else wrote the introduction, his name would be designated next. Finally, there is an imprint, showing the publisher and the place and date of publication. The back side of the title page usually gives the date of copyright, the names of the copyright owners, and the Library of Congress number.

The *preface* or *foreword* follows the title page. It is here that the author states his purpose in writing the book. He tells for whom the book is intended, and perhaps why. He expresses gratitude to those who have materially assisted in writing the book. He might also describe any special features of the book.

The *contents* lists the chapters of the book, with chapter titles, and the page on which the chapter begins. It may also list section titles and perhaps their pages. The table of contents serves as an outline of the book. If you are looking for some major topic, a quick glance through the table of contents can show whether a particular book covers that topic.

A list of *illustrations* following the table of contents is sometimes included. This list is presented in the order that the illustrations appear in the text. The list is useful in helping you locate one of the illustrations quickly.

The *introduction* to a book is different from the preface in several respects. It concerns the subject matter of the book and how the material is presented. The introduction prepares the reader for the content of the book.

The *text* is the main body of subject matter being presented in the book. In scientific and technical materials, it is here that the author explains laws or principles, describes physical systems, or makes his definitions. In textbooks, the *text* may include end of section, or end of chapter, problems. Example problems are frequently used. Illustrations, figures, diagrams, and graphs appear to support discussion in the text.

Following the text of a book, auxiliary material may include an *appendix*, *bibliography*, *glossary*, and *index*. The *appendix* is used in technical books to contain tables, like those for trigonometric ratios or logarithms. Any facts, data, or notes not easily included in the text, but useful to the student, may be found in an appendix. The bibliography could be placed at the end of a book, but some bibliographies are placed at the ends of chapters. A *bibliography* is a list of books, specific articles in periodicals, manuscripts, documents, etc., relating directly to the subject being discussed. In a scientific or technical textbook, the bibliography might include journal articles reporting the original discovery of some principle or device being discussed or books used as references in preparing the text. Sometimes bibliographies are used to indicate more advanced or extensive material on some subject. The *glossary* lists and explains technical or foreign words used in the text but not explained there.

One of the most important parts of a book is the *index*. Without an index, the specific information you want from a book may be hopelessly buried within those hundreds of pages. The index provides you with a direct route to any specific item of importance in the book. It is indeed strange that people seem to use the index as a last resort for finding information when "thumbing" through the pages leads nowhere. Failing to use a book index is almost as bad as looking for someone's telephone number by thumbing through the telephone book, looking at the addresses until you find the right one!

The *index* is a list of specific items of importance discussed in the book. It is arranged alphabetically, and includes all topics, people, and places mentioned in the text. Each index entry is followed by the number of the page or pages on which the subject is discussed. For example, Figure 13-1 shows an index page from a physical science textbook. Notice that some subjects have sub-entries. The third sub-entry under Energy is, "kinetic, 179, Sec. 9-5." Not only does this entry refer to kinetic energy, which is discussed on page 179 of that text, but the words kinetic energy must be included in the title of Sec. 9-5.

We have, in this section, discussed briefly the parts of a book. Your ability to use a textbook effectively will depend to a great extent upon your understanding of the functions of these parts.

Index

A page number in boldface type indicates that the item appears in boldface on that page. A section number usually means extended discussion of the item. When an item is most easily located from a figure caption, the figure number is listed.

Acceleration, Sec. 9-2
   average, **53**, Sec. 9-2
   gravity, **176**, Secs. 9-10, 9-11, 9-12
Accuracy, **102**
   per cent, Sec. 5-4
Acor, Sec. 6-3
Action, Sec. 6-2, **92**
   at a distance, Sec. 4-1
Articulated science teaching, 11
Alchemy, 3
Archimedes, 1, 14, 192
Archimedes' principle, Sec. 8-7
Associative law, **56**
Astin, Dr. Allen V., 69
Attribute, **45**, Secs. 6-2, 6-3
   invariant, **45**, Sec. 3-5
   conaddian, **47**
   key, **92**, Secs. 6-2, 6-3

Balanced emphasis, 11
Binary, **70**
Boldface type, 2
Brahe, 194
BTU, 184 (Table), 185
Buoyancy, Sec. 8-7

Cal, 185
Calculus, **176**
Center of gravity, 128
Center of mass, 190
CG, 128, 190
CG — weight point, Sec. 7-10
Centrifugal force, **190**
Centripetal force, **190**
Chaos, 122
Coefficient of friction, Fig. 9-4
Commutative law, **55, 111**
Comparison operation, 45, 76, Sec. 6-3
Component of vector, **109**
Compression, 89
Conaddian, 47

Conceptual experiments, 159, *(see also* Experiments)
Conservation, **8**
   angular momentum, **191**
   atoms, 3, 7
   continuous, 8
   energy, 3, 7, 16, Ch. 8, Ch. 9
   in general, 186-191
   intermittent, **8, 180**
   laws, **187, 191**
   momentum, **177**
   objects, 48
   potential energy, 163, Secs. 8-5, 8-6
Coordinates, cartesian, 57
Creative search pattern, 16, **18**, Sec. 2-13, **122**
CSP, 122

Deformed, **86**
Descartes, 55, 56
Diagrams, 57, Sec. 6-2
Dimensional analysis, **183**

Einstein, 62, 178
Efficiency, **185**
Ellipse, 195
Energy, Ch. 1, Sec. 9-4, QLF 9-3
   electromagnetic, Sec. 9-6
   heat, Sec. 9-6
   kinetic, **179**, Sec. 9-5
   mass-energy, QLF 3-4
   potential, Sec. 8-4
Equation:
   NSU, **51**, 76
   qualified, **30**, 43, Sec. 6-3
   qualifier, **30**
Equilibrium:
   neutral, **141, 187**
   stable, **140, 187**
   unstable, **140, 187**
Erg, 184 (Table), 185

Figure 13-1. *An index from a physical science textbook.**

---

*Shockley, William and Gong, Walter, *Mechanics* (Columbus: Charles E. Merrill Books, Inc., 1966), p. 211.

PROBLEMS

1. Examine each part of this textbook, listing them in the actual order in which they appear. Make a similar list for books of each course in which you are now enrolled.

2. From Figure 13-1, locate the page numbers for each of the following specific subjects.

    a. component of a vector   b. potential energy

    c. the law of conservation of momentum

    d. the efficiency of a machine

13-3 The Library Catalog System

Knowing the parts of a book and their function helps you use books to get needed information. The preface tells you whether the book is at your level and whether it contains the kind of material you are looking for. If you are searching for information on some major area, the table of contents can reveal its presence in a book. If you are looking for some specific topic, the index will locate the page number immediately. Other parts of a book can be used as needed, but only if you are aware of their existence.

Being able to use a book effectively is important, but that skill is wasted if you use it only with your required textbooks. A textbook is written for the average student. You will find parts of it easy for you and parts of it quite difficult, or even incomprehensible. What is easy for you might be quite difficult for someone else. It depends upon your background and experience. Very often you need to refer to some other book at a lower level; or perhaps when you find something that appears particularly interesting, you would like to read more extensively about it. The library has books, periodicals, and other materials of almost every kind and level. They are there for you to use. All you have to do is find what you want.

All you have to do is find it. That statement is easy to make, but few people are efficient in using a library. Indeed, some would not even know how to begin looking for a book without assistance, much less know how to look for periodical literature. They search for information in a library much like they search a book without using the index. They "thumb" through the library, hoping to stumble onto that one reference for which they are looking. It is not surprising that they consider it "too much trouble" to look up information themselves in the library.

How do we find information in a library? Before answering this question, let us examine how books are arranged and how they are classified. If you understand the classification system, you will then understand how books are arranged. Later we will discuss how information is located.

Because of the convenience to the reader in keeping together books on the same subject, the classification of books is in terms of *subject*. Although there are a number of classification systems used in various libraries, the most common is the *Dewey Decimal System*. This system assigns a number to each book, the numbers corresponding to certain subject areas. The following brief list shows the major Dewey decimal classifications.

- 000 General works
- 100 Philosophy
- 200 Religion
- 300 Social sciences
- 400 Languages
- 500 Pure science
- 600 Useful arts or applied science
- 700 Fine arts, Recreation
- 800 Literature
- 900 History

Each of these major classifications is further broken down into sub-classifications. For example, in the area of 500 pure science, there are sub-classifications as follows:

- 500 Pure science
- 510 Mathematics
- 520 Astronomy
- 530 Physics
- 540 Chemistry and allied sciences
- 550 Earth sciences
- 560 Paleontology
- 570 Anthropology and biology
- 580 Botanical sciences
- 590 Zoological sciences

Besides the ones listed above, a person studying science or technology should be familiar with at least the following sub-classifications:

- 600 Technology
- 610 Medical sciences
- 620 Engineering
- 660 Chemical technology

Using this classification system, books are assigned *call numbers*. The call number consists of the Dewey decimal classification of the book, below which is placed a *book number* or *author number* to further differentiate the book from others on the same subject. For example, we might have a physics book with the following call number:

532.4
B21

The author's last name begins with B; the number 21 is assigned through use of a table which places authors' names in alphabetical order. We need not be concerned here with how these assignments are made.

Books are arranged on the shelves of libraries according to the call numbers, in numerical order for the Dewey decimal classification, and when identical numbers appear, alphabetically for the author or book number. Displayed on the end of each stack are labels indicating call numbers located in that particular stack. Larger libraries usually have classification directories for locating subject areas.

In many libraries, fiction and biography are not classified. Instead, where the Dewey decimal number would appear, the letter F (or Fic) may appear for fiction. B is sometimes used to indicate a biography. Below this letter will appear an author symbol, either the first letter of his last name, or an author number, as used in other call numbers.

In any given library there will be variations in the system of assigning these call numbers or other classifications. You should determine what special methods are used in your own library, but the understanding of the classification system just discussed will enable you to locate most books you will need.

Being familiar with the classification system used in your library is only a first step in learning to use it. If the library is small, it might not be too difficult to locate most needed subject areas by walking around looking for major classification numbers. But you may want to use a large library, like the specialized *Linda Hall Library* on the campus of The University of Missouri, at Kansas City, Missouri. This library has books with classifications only in the 500's and 600's, and it has excluded the biological sciences. It is a scientific and technical library. The physics section alone contains several thousand volumes. You would need to know more than the sub-classification number 530 to find a given physics book in that collection. You must learn to locate books using the library catalog system.

A library catalog system is an index to the library's holdings filed alphabetically by subject, author, and title. The catalog may be in card or book form, but the basic techniques for using it are the same.

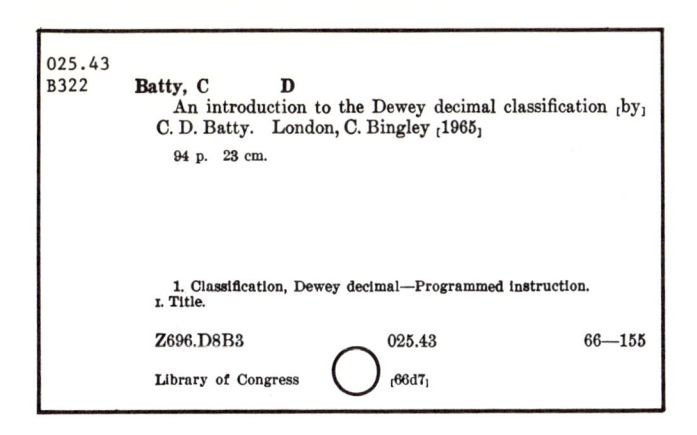

```
025.43
B322    Batty, C        D
            An introduction to the Dewey decimal classification [by]
            C. D. Batty. London, C. Bingley [1965]
            94 p. 23 cm.




                1. Classification, Dewey decimal—Programmed Instruction.
                I. Title.

            Z696.D8B3                   025.43              66—155
            Library of Congress       ( )  [66d7]
```

```
                    An Introduction to the Dewey Decimal
025.43          Classification
B322    Batty, C        D
            An introduction to the Dewey decimal classification [by]
            C. D. Batty. London, C. Bingley [1965]
            94 p. 23 cm.




                1. Classification, Dewey decimal—Programmed Instruction.
                I. Title.

            Z696.D8B3       ( )   025.43              66—155
            Library of Congress       [66d7]
```

```
025.43
B322       CLASSIFICATION, DEWEY DECIMAL--PROGRAMMED
        Batty, C       D,                  INSTRUCTION
            An introduction to the Dewey decimal classification [by]
            C. D. Batty. London, C. Bingley [1965]
            94 p. 23 cm.




                1. Classification, Dewey decimal—Programmed instruction.
                I. Title.

            Z696.D8B3       ( )   025.43              66—155
            Library of Congress       [66d7]
```

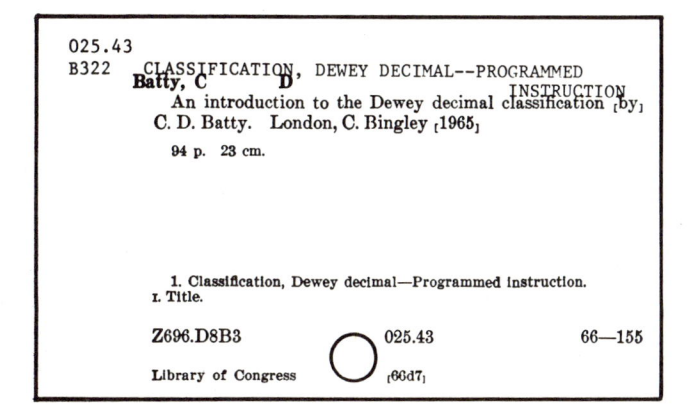

Figure 13-2.   *Three basic cards from a library catalog.*

Most books have at least three entries in the catalog system; the author card, the subject card, and title card. The author card is the main entry in the catalog system for a given book. It indicates responsibility of authorship, whether it belongs to a person, a periodical, a corporation, or a university. It includes the title and sub-title, the edition, publication information, and the physical description of the book (the number of pages or volumes, the size of the book in centimeters, illustrations, bibliography, etc.). Most important, the card contains the call number of the book in the upper left hand corner. This number is used to locate the book in your library.

A title card is a reproduction of the author card, but typed above the author's name is the title of the book. A subject card is made in a similar way, except that above the author's name the subject of the book is typed in capital letters. Sometimes this subject is typed in red letters. Figure 13-2 illustrates the three basic kinds of cards: author, subject, and title.

This brief discussion of the library catalog system should help you to locate books in your own library. You will find your librarian most cooperative in assisting you in using the catalog. As a student of science or engineering, your primary interest may be in books of the 500 or 600 classification, but you should use the catalog to locate specific book titles, subjects, or authors.

PROBLEMS

1. In what major classification in the Dewey decimal system would you find each of the following book titles?
    a. *Calculus and Analytic Geometry*
    b. *History of Physics*
    c. *Metaphysics*
    d. *The Chemistry of Inorganic substances*

2. In what major classification would you find books with the following call numbers?
    a. 808.5   b. 029.6   c. 901
        G792b2   T929s   L691

3. In what sub-classification would you find books with the following call numbers?
    a. 512.3   b. 608.1   c. 627.32
        B13   D1   C11

4. Where would you find books with the following call numbers?
    a. F   b. 571.8   c. B
        R894   G112   A2

13-4 Library Materials in Science and Technology

We have discussed textbooks, the parts of a book, and the library catalog system. A knowledge of these topics can help you locate material in a library when you need it. Of course, a library contains many more sources of information than just books. It also contains periodical literature and pamphlets. More important, it contains reference material and indexes which help you locate information of a specific kind.

Most of the kinds of materials needed for science or technology courses can be found in special reference books, periodicals, or regular non-fiction, technical books. If you want more information about some subject in a technical course, and you want to look first at books, use the subject index of your library catalog. Locate the specific numbers appropriate to that subject, then go to that part of the stacks, or, if you do not have stack access, ask for specific call numbers. The catalog will give you sufficient information about the contents of books to justify your getting three or four of them to examine.

When you have located a book on the subject of interest, use the table of contents to locate a general topic, or, if you have a specific topic, use the index to locate the page number. You may want to read the preface first to see if the book is at a level you can understand or if it is too elementary. When you find what you want, take notes just as you would from your own textbook.

Very often you may want to get information on some new application in science or technology. Such information is usually found in a periodical. It is exciting to go back to the original journal article written by some famous scientist or engineer. If you can follow what he says, you get first-hand what you often find diluted in your textbook. For example, the original article in *Nature* by James Chadwick on his discovery of the neutron is short and easy reading. For the interested reader, the article is titled, "Possible Existence of a Neutron", and appears on page 312 of Volume 129 of *Nature*, dated February 27, 1932.

There are many indexes of periodical literature. These indexes are organized alphabetically by subject and author of article, and appear for each year. The most recent indexes usually appear in weekly and monthly forms. If you are looking for information on lasers, you will find many entries with specific titles in various periodicals. By jotting down the journal, the volume, date, page, and article title, you can check the library catalog system to locate the particular article.

For general periodicals, the most common index is the *Reader's Guide to Periodical Literature*. Virtually every library will have this index. The most useful index for students of science and technology is the *Applied Science and Technology Index*. This index is one of two which replaced the *Industrial Arts Index* (1913-1957). It is a subject index to periodicals in the fields of aeronautics, automation, physics, chemistry, engineering, industrial and mechanical arts, and related fields.

The index to periodical literature enables you to locate periodical information on any subject. There are, of course, many other kinds of reference sources in a library. Specialized reference books, dictionaries, handbooks, and encyclopedias can be helpful at times. You should explore your own library to see what specific reference volumes are available.

This chapter has given you only a brief and limited description of how to find library information in science and technology. You must learn to use a library, and the topics discussed here should help get you started in learning about yours. You will find that when you can use it, the library will be your most essential and dependable source of information for the rest of your life. When you learn to use it, you will no longer need teachers. You can learn anything to which you are willing to give sufficient time and energy, and the library will teach it to you.

In science and technical fields, there is one source of information not yet mentioned here. It is the one source which is superior to all others. It is the ultimate authority, the appeal to nature herself. You are making that appeal when you carry out a scientific measurement or experiment. No matter what is said in a book, if nature does not agree, we change the book. Science is not based ultimately on what *people* say or have said in the past. It is not an authoritarian system. It is based on what we can confirm as we examine nature through measurement. It is for this reason that science must be quantitative.

PROBLEMS

1. Go to your own college library and locate the following references:
    a. *Chamber's Technical Dictionary*
    b. *The McGraw-Hill Encyclopedia of Science and Technology*
    c. *Handbook of Chemistry and Physics*
    d. *Van Nostrand's Scientific Encyclopedia*

2. Go to the indexes in your college library and make a list of the most recent periodicals having information on the following subjects.
    a. Gas lasers   b. Tunnel diodes   c. Lung cancer   d. Nuclear reactors

Answers to Selected Problems

Section 1-1

1. The ratio L/W is the same for any unit distance.

3. 0.5

Section 1-4

1. 12,000,000,000 dollars

3. 0.0006 T-people

5. 5,000,000 nsec

7. 31,556,925,975,456 μsec

Section 2-1
1. a. The number to be added to 3 to give 5 is   c. The number to be added to 12 to give 4 is   e. The number to be added to X to give 3 is
2. a. no solution   c. 135   e. no solution   g. 256

Section 2-2
1. a. 54   c. 4   e. 512
2. a. -4 c. -12.32

Section 2-3
1. a. 11/4   c. no solution
2. a. 1/5   c. 10/3

Section 2-8
1. a. -3.5 c. 4.4   e. -1.5
2. a. -15 c. 13.02   e. 4

Section 2-9
1. represents π

Section 3-1
1. a. 10^3 m   c. 10^{12} sec
2. a. 10^{10} c. 10^9
3. first difference = 9000, second difference = 90,000,000,000,000 mi

Section 3-2
1. a. 10^{11} c. 10^9
2. a. 10^7 c. 10^{21}
3. a. 10,000,000   c. 1000

Section 3-3

1. a. 10^5 c. 10^{18}
2. a. 10^2 c. 10^6

Section 3-4

1. a. 10^{-1} c. 10^{-3}
2. a. 10^{-6} m   c. 10^{-3} m
3. a. 0.0000001   c. 0.01

Section 3-5

1. a. 1   c. $\frac{1}{2}$
2. a. 1   c. 1

Section 3-6

1. 3.14×10^{11} m
3. 0.000000000000000000160
5. a. 1.36×10^{-6} c. 2.00056×10^{-3}

Section 3-7

1. a. 3×10^{91} c. 5.6×10^9
2. a. 3.6×10^{-9} c. 2×10^{-27} e. 1.6×10^3

Section 3-8

1. a. 2×10^2 c. 4.5×10^{25}
2. a. 2   c. 2×10^{16}
3. 6.25×10^{-7}

Section 3-9

1. a. 4.9×10^{-10} c. 6.4×10^4 e. 3.6×10^{-18}
2. a. 6×10^7 c. 2×10^{-12}

3. a. 7×10^2 c. 8×10^{-7}

Section 4-3
1. a. 268 c. 592
2. a. 206 c. 722

Section 4-4
1. a. 1.01×10^8 c. 3.46×10^{-6} e. 1.78×10^{-10}
2. a. 1.98×10^{-1} c. 3.44×10^{-14}

Section 4-5
1. a. 2.46×10^1 c. 4.18×10^1 e. 1.45×10^8 g. 1.53×10^{-1}

Section 4-6
1. a. 6.43×10^{10} c. 3.16×10^{-2}
2. a. 2.5×10^{-9}

Section 4-7
1. a. 1.19×10^{23} c. 2.96×10^{-65}
2. a. 9.87×10^5 c. 6.66×10^{-1} e. 5.72×10^{-2}
    g. 2.57×10^{-2} i. 6.95×10^{-8} k. 7.02×10^{-8}
    m. 9.9×10^2

Section 5-1
1. a. S, V, etc. c. V, τ, etc.
2. b. $\frac{1}{2}\rho V^2 + \rho g h + P = C$

Section 5-2
1. b. The velocity at time T is equivalent to the difference of the initial velocity and the product of the acceleration of gravity and the time required.

Answers to Selected Problems

2. a. 18 c. 2 e. 37.5

3. a. $V =$ c. $P - 0.03P$ e. $M = 3S - 12$

Section 5-3

1. a. $-107/36$ c. $38/3$ e. $5/144$

2. a. 3 c. 5 e. 6

Section 5-4

1. a. conditional c. identical

2. a. $X^2 - 5 = -1$ c. $Y^2/X^2 - 5 = 20$

Section 5-5

1. a. no solution c. no solution e. $X = 5$
g. no solution i. $X = 8$

2. a. no solution c. $X = -4$ e. $X = 1$ g. no solution

3. a. $X = 12.6$ c. no solution e. $X = -4$ g. $X = -2, 2$

4. a. -1 c. 8 e. $-215/6$

5. a. (1) square X; (2) multiply (1) by 3; (3) subtract 4 from (2); (4) divide (3) by 2; (5) add 5 to (4)
c. (1) subtract B from X; (2) multiply (1) by A; (3) divide (2) by $(C - D)$; (4) add E to (3)

6. a. $a = F/m$ c. $x = (V^2 - V_0^2)/2a$ e. $M = uv/a + ut$
g. $h = \omega^2(mr^2 + J)/2u$ i. $g = E/(1 - R/r)mR$

Section 6-1

1. 39.5

3. 0.5; 0.866

4. a. 2.0 c. 0.682 e. 15

5. 1.79×10^{-25} kg

260 *Quantitative Aspects of Science and Technology*

Section 6-2

1. 10

3. 1/3 as great; V is proportional to T

5. 4

Section 6-3

1. R is proportional to V; 15; $R = 15V$

3. a. direct; 4; $Y = 4X$ c. no simple relationship
e. inverse; 100; $X = 100/Y$

Section 6-4

1. $\overline{EF} = 36$; $\overline{FD} = 28$

3. a. 0.33   c. 0.577   e. 0.5

Section 6-5

1. $7.96 \times 10^{23} \text{m}^3$; $7.15 \times 10^2 \text{ kg/m}^3$

3. Hint: Show that $F/a = $ constant. Hint: Divide equation $F_1 = ma_1$ by $F_2 = ma_2$.

5. E is proportional to the square of R, q, and B; E is inversely proportional to m; E increases by $3^2 = 9$.

7. a. 1. increased by a factor of two   2. decreased by a factor of two   3. decreased by a factor of four   4. increased by a factor of four   5. decreased by a factor of eight
c. 1. increased by a factor of three   2. decreased by a factor of three   3. decreased by a factor of nine   4. increased by a factor of nine   5. decreased by a factor of twenty-seven

Section 6-6

1. 10^{-4} as much; 10^{-6} as much; loses heat too fast to keep warm

3. 10^4; 10^6; make the gear linear dimension at least 1000 times that of the model

Section 7-1

1. a. 3.4873 c. 9012.58

3. a. 0.9144 c. 1610 e. 2.0368

Section 7-2

1. a. 12.3 mm c. 16.8 mm

2. a. 57/64 c. 13/32

Section 7-3

1. a. 17 g c. 226 g

2. a. 53.37 g c. 27.52 g

3. 34.98 g

5. 8.0

Section 7-4

1. a. 3,628,800 sec c. 357,732.3

3. 10^5

Section 7-5

1. 10^{-5} m

3. 0.05 sec

5. decreased slightly

Section 7-6

1. 0.5; 0.0278

3. 6.67×10^{-4}

5. 0.054

Section 8-1

1. a; c; d; f; h
3. a. L, T c. L, T e. Q, T

Section 8-2

1. T independent; V dependent; P held fixed
3. I independent; V dependent; $V = 2I$

Section 8-4

1. 169.5
3. 74.356

Section 8-5

1. a. 4   c. 9
3. 0.118; 0.141; 0.0262; 0.0313
5. 9.836 ma; 0.023 ma; 0.00234

Section 8-6

1. $\bar{L} = 51$; $S = 15.5$; $S_{\bar{x}} = 4.9$; 46 to 56; 35 to 67
3. $\bar{T} = 0.9958$; $\bar{H} = 16.031$; $\delta \bar{T} = \pm\, 0.00045$; $\delta \bar{H} = \pm\, 0.0068$; $R_{\bar{T}} = 4.5 \times 10^{-4}$; $R_{\bar{H}} = 4.2 \times 10^{-4}$

Section 8-7

1. (1) $PE = 10.4$; $PE_{\bar{L}} = 3.3$ (3) T: $PE = 0.0012$; $PE_{\bar{T}} = 0.00030$ H: $PE = 0.018$; $PE_{\bar{H}} = 0.0046$

Section 9-3

3. slope = 0.492 v/amp; intercept = 0
5. a. $-5; 6$ c. $2; -9$ e. $4/5; -3/5$

Section 9-4

1. $P = 21,450/V$

3. Use Y versus X^2, slope -5, intercept 8; $Y = -5X^2 + 8$

Section 9-5

1. a. $\log_6 Y = X$ c. $\log_5 25 = 2$ e. $\log_{10} 100,000 = 5$
2. a. $b^Y = X$ c. $5^4 = 625$ e. $12^0 = 1$
3. a. 3   c. 4   e. 5

Section 9-6

1. slope $= 1/2$; intercept $= 2.03$; $T = 2.03\sqrt{L}$
3. slope $= -2$; intercept $= 1000$; $F = 1000/R^2$

Section 10-1

1. torsion constant⟷stiffness of spring; rotational inertia⟷mass on spring; angle of twist⟷amplitude of vibration

Section 10-2

1. 2.5 p/in.
3. 1.6 sec

Section 10-5

1. slope $= -1/2$; intercept $= 6.28$; $T = 6.28/\sqrt{g}$; $T = 6.28\sqrt{L/g}$

Section 10-6

1. 1.35×10^5

Section 11-1

1. 450 units
3. $MV^2/2\tau$

Section 11-3

1. $n = m/M$; $PV = (m/M)RT$; $PV/RT = m/M$; $MPV/RT = m$; $MP/RT = m/V$

3. a. $T^2 = 4\pi^2 L/g$; $gT^2 = 4\pi^2 L$; $g = 4\pi^2 L/T^2$

Section 12-1

1. Given: $r = 24$ in.; $\Delta r = 0.02$ in.; $T = 20°C$; Find: ΔT

Section 12-2

1. a. Given: $F = 30$ n; $A = 5$ m/sec^2; Find: M; Relationship: $F = MA$ c. Given: $T = 10$ sec; $A = 8$ m/sec^2; $V_i = 4$ m/sec; Find: X; Relationship: $V_i T + \frac{1}{2}AT^2 = X$

Section 12-3

1. $X = [(V_f + V_i)/2]T$; $X = (V_f T + V_i T)/2$; $X = [(V_i + AT)T + V_i T]/2$; $X = (V_i T + AT^2 + V_i T)/2$; $X = (2V_i T + AT^2)/2$; $X = V_i T + (\frac{1}{2})AT^2$

Section 12-4

1. $P = 25.15$; $\delta P = 1.78$

3. 77 newtons

Section 12-5

1. Given: $w = 16$ p; $h = 64$ ft; Find: U; Relationship: $U = wh$; $U = (16)(64) = 1024$ ft − p

Section 13-3

1. a. 510   c. 100

3. a. Mathematics   c. Engineering

Index

Abbreviations for units, 7
Abscissa (*see* Graphs)
Absolute uncertainty
  (*see* Uncertainties)
Addition
  familiar process, 17
  natural numbers, 11
  rules, 18, 22
Algebra, fundamental principle, 72
Algebraic symbols, 64–66
Amplitude of vibration, 195
  (*see also* Experiment)
Appendix of a book (*see* Book)
Applied Science and Technology Index, 253
Arithmetic mean, 141–144
  (*see also* Mean)
Atomic clock, 122
Author number
  (*see* Dewey Decimal System)

Balance
  equal arm, 114–115
    (*see also* Mass)
  triple beam, 116–117
    (*see also* Mass)
  unequal arm, 115–120
    (*see also* Mass)
Bibliography of a book (*see* Book)
Book number
  (*see* Dewey Decimal System)
Book
  appendix, 246
  bibliography, 246
  contents, 245
  foreword, 245
  glossary, 246
  illustrations, 245
  index, 246–247
  introduction, 245
  parts of, 245–248
  preface, 245
  text, 245

Book (continued)
  title page, 245
Boyle's law, 97
Brahe, Tycho, 80
Bureau of Standards
  time broadcasts, 125

Calibration of instruments, 123–128
Catalog system of library, 248–252
  (see also Dewey Decimal System)
Chadwick, James, 253
Charge, 4
Charles' law, 100
Circle
  circumference, 23
  ratio of C to D, 82
Computations
  general, 42–43 (see also Slide rule)
  in problem solving, 236–240
  uncertainties in, 238–239
Constant of proportionality, 85
Contents of a book (see Book)
Coordinate axes (see Graphs)
Cube root, 42

Data analysis
  mass on a spring (see Experiment)
  uncertainties, 200–203
  (see also Uncertainties)
Data sheets, 138–139
Deductions in reading, 219–223
Definitions
  operational, 4–5
  reading of, 213–216
  stating, 213–215
Density, 96
Derivations
  assumptions in, 222
  reading, 219–223
Deviation, 146
Dewey Decimal System
  author number, 250
  book number, 250
  major classifications, 249
  technical sub-classifications, 249
Diagrams, reading, 216–219

Distribution
  Gaussian, 143
  sample, 148–149
Dividend, 55
Division
  definition, 21
  indicated, 67
  large and small numbers, 39–41
  with powers of ten, 29–30, 39–40
  rules
    for base ten, 31
    of exponents, 30, 34
    for signed numbers, 21–22
  slide rule, 54–57
  symbol, 21, 29
Divisor, 55

Earth
  mass, 36
  orbital radius, 36
Einstein, Albert, 113
Electron
  charge, 36
  mass, 81
  ratio charge to mass, 81
English units, length, 104–108
Equal sign, 67
Equations
  conditional, 71
  identical, 71
  mathematical, 71–72
  operations on, 71–72
  parts of, 71
  from proportions, 96–98
  solving algebraic, 72–77
Equivalence, 71
Experiment
  distinguished from measurement, 135
  mass on a spring
    conclusions, 209–211
    data analysis, 200–204
    design, 193–196
    equation, 207–208
    identification of variables, 191–193

Experiment (continued)
  mass on a spring (continued)
    period, 210
    recording data, 137–141
    variables, description of, 136
Experimental method, 134–137
Exponents
  fractional, 41–42
  negative, 32, 34
  positive integral, 26
  rules
    division, 30, 179
    multiplication, 28, 179
    powers, 41, 179
    zero, 33–34
Extraneous variables, 138

Factor, 26
Figures, reading of, 216–219
Foot, 6
Foreword of a book (*see* Book)
Fraction, 15
Fractional exponents
  (*see* Exponents)
Functions, non-linear, 174–178
Fundamental physical quantities,
  2–4 (*see also* Physical quantities)

Giants, impossibility of, 101
Glossary of a book (*see* Book)
Graphs
  abscissa, 162
  coordinate axes, 160–163
  error indications on, 168
  horizontal axis, 160
  ordinate, 163
  origin, 160
  of points, 160–169
  reading, 216–217
  vertical axis, 160
  vertical intercept, 171
Gravity, acceleration of, 112
Greek symbols, 66

Harmonic motion, 191
Hooke's law, 96

Horizontal axis (*see* Graphs)
Hypotenuse, 92

Identity
  additive, 14
  as equation, 71
  multiplicative, 15
Illustrations list in books (*see* Book)
Index of a book (*see* Book)
Indicated product, 67
Industrial Arts Index, 254
Inertia (*see* Mass)
Inertial mass (*see* Mass)
Integers
  definition, 26
  description, 14
Introduction of a book (*see* Book)
Inverse
  additive, 14
  of indicated operations, 73
  multiplicative, 15
Inverse square law
  for light, 100
  as a proportion, 97–99
Irrational numbers, 23–24
Italics, 214

Kepler, Johannes, 80, 187
Kepler's third law
  (*see* Planetary motion)
Kilogram subdivisions, 6
  (*see also* Mass)
Kinetic energy, 214

Legs of right triangles, 92–93
Length
  measurements of, 104–108
  units of, 3, 6, 124
Library catalog system, 248–252
  (*see also* Dewey Decimal System)
Light, speed of, 114
Linda Hall Library, 250
Linear dimension in scaling laws, 101
Linear functions
  algebraic form, 169
  graphs of, 169–174

Linear functions (continued)
  slope-intercept form, 174
Linear relationships, 164
Logarithms
  definition, 179
  rules for, 180
Log-log graph
  method of use, 189
  plots, 186–189
  slope of, 186
  straight lines on, 206
  vertical intercept, 186

Magnitude, order of, 28
Mass
  atomic, 114
  comparisons with balance, 114–120
  confused with weight, 3
  equivalence, gravitational and inertial, 4, 113
  gravitational, 3, 113–114
  inertial, 3, 113
  international unit, 6
  measurements of, 112–120
  preparation of standard, 125
  on spring (*see* Experiment)
Mean
  as best estimate, 143
  definition, 143
  standard deviation of, 154–156
Mean deviation
  definition, 146
  as error estimate, 147
  example, 147
  for measurements, 144–148
  as measure of spread, 146
Measurement
  derived, 103
  distinguished from experiment, 135
  uncertainties in, 128–131
  (*see also* Uncertainties)
Measuring instruments, calibration
  (*see* Calibration)
Meter
  subdivisions, 6
  as a unit, 3

Metric units of length, 104–108
Michelson interferometer, 124
Micro-balance, 117–118
Minus sign, 67
MKS system, fundamental units, 7
Multiplication
  indicated, 67
  large and small numbers, 37–39
  with powers of ten, 28, 37–39
  rational numbers, 20–22
  rules
    base ten, 31
    with exponents, 28, 34
  slide rule, 51–53
  symbols for, 26

Natural numbers, 10–12
Negative, definition of, 26
Negative exponents, 31–33
Negative numbers, 13–14
Negative sign, subtraction, 12
Newton, Isaac, 113
Non-linear functions, 174–178
Numbers
  complex, 24
  integral, 13–14
  irrational, 23
  natural, 10–11
  power of ten form, 34–36
  rational, 15–22
  real, 23
  rules of operation, 22
Numeral, 10

Operational process, 4–5
Operations
  indicated in equations, 73
  rules, 22
Operators, mathematical, 70
Origin of graph (*see* Graphs)
Ordinate (*see* Graphs)
Oscilloscope, 126

Parentheses for multiplication, 27
Pendulum, 209
Per cent error, 130–131

Period of vibration, 194
Physical definitions, 4–5
Physical quantities
  derived, 133–134
  fundamental, 2, 6–8
Pi, 23
Planck's constant, 65
Planetary motion law, 187–188
Plus sign, 26, 67
Positive, 26
Powers, rule for, 41
Powers of ten
  division, 29–30
  multiplication, 28–29
  numbers, 34–36
  rules for expression, 35–36
Precision
  dependence on object, 105
  instrument error, 129–130
Preface of a book (see Book)
Prefixes
  numerical, 8
  table, 8
  for units, 7
Pressure, 5
Probable error
  as estimate of uncertainty, 157–158
  formula, 157
  meaning, 157–158
Probable error of mean, 157–158
Problems (see Word problems)
Proportion
  common types, 98
  conversion to equation, 97
  definition, 83, 86
  direct
    definition, 83–84
    example, 86–87
  general definition, 84
  inverse
    definition, 84
    example, 87–88
  symbol, 83
Proportional sides in triangles, 93–94
Proportionality
  constant of, 85–88

Proportionality (continued)
  meaning of, 82–85
  test for, 86
Pythagorean theorem, 93

Quotient, 55

Radioactive half-life, 185
Range, 142
Ratio
  charge to mass, electron, 81
  for comparisons, 79–80
  definition, 80
  meaning of, 79–82
Ratios
  planet-sun to earth-sun distances, 80
  trigonometric, 94
Rational numbers
  addition, 16–18
  division, 21–22
  subtraction, 19–20
  summary of rules for, 22
Reader's Guide to
  Periodical Literature, 253
Reciprocals, 15
Relationships, deducing, 233–235
Relative uncertainty, 130
  (see also Uncertainties)
Right triangle, 92
Roots and powers, 41–42

Scale factors, 100–102
Scale models, 102
Scaling laws, 100–102
Second
  definition, 122
  as a unit, 3
Semi-logarithmic graphs
  plots, 179–185
  slope, 181
  vertical intercept, 181
Signed numbers
  (see Rational numbers)
Significant figures, 237
Similarity, 89

Similar triangles, 89–95
Slide rule
  division, 54–57
  multiple operations, 57–59
  multiplication, 51–53
  parts, 44–45
  scales, 45–50
  square roots, 59–62
  squares, 62–63
Slope, 171, 181, 186
  (see also Graphs)
Slug, as a mass unit, 112
Solve, meaning of, 72
Space measure, 2
Speed, operational definition, 5
Speedometer calibration, 133–134
Spring constant
  definition, 194
  as slope of graph, 201
  (see also Experiment)
Square root
  related to exponent ½, 41
  symbol as operator, 70
Standard deviation
  formula, 148
  meaning of, 150–154
  for measurements, 148–156
Standard deviation of mean
  definition, 154
  as error estimate, 155–156
  example of, 155–156
  formula, 154
Standard length, 6
Standard mass, 6
Strength of structures, 101
Studying, definitions, 224
Studying, how to do it, 223–225
Studying, steps needed, 225
Studying, taking notes, 224
Subdivisions, slide rule scales, 47–49
Subtraction
  definition, 11–12, 19
  natural numbers, 11
  rule, 20, 22
Symbols
  definition, 10

Symbols (continued)
  for physical quantities, 64–65
System
  English units, 6
  metric units, 6
  (see also MKS system)

Tables on data sheets, 138–139
  (see also Experiment)
Text of a book (see Book)
Textbooks, 244
Theory of mass on spring, 210
  (see also Experiment)
Time interval
  measurement, 120–123
  standard unit, 122
Title page of book (see Book)
Triangles
  corresponding sides, 90
  right, 92
Trigonometric ratios (see Ratios)

Uncertainties
  absolute, 129–130
  propagation of, 236–237
Units of length
  English, 104–108
  English to metric, 107–108
  metric, 104–108
Universal constants, 65
Universal gravitation, 65, 97–98

Verbal statements to symbolic form, 67–69
Vernier index, 109
Vernier scale
  description, 108–109
  reading of, 110–112
Vertical axis (see Graphs)
Vertical intercept (see Graphs)
Vibrating system variables, 191–193
  (see also Experiment)
Voltmeter, digital, 127

Weight
  as force, 112

Weight (continued)
  of structures, 101
Word problems
  deducing relationships, 233–235
  plug-in type, 234
  reading and analysis, 227–231

Word problems (continued)
  relationships, 231–233
  systematic solution, 240
Year, as time unit, 3
Zero exponent, 33–34
  (*see also* Exponents)